Citizenship and Infrastructure

This book brings together insights from leading urban scholars and explicitly develops the connections between infrastructure and citizenship.

It demonstrates the ways in which adopting an 'infrastructural citizenship' lens illuminates a broader understanding of the material and civic nature of urban life for both citizens and the state. Drawing on examples of housing, water, electricity and sanitation across Africa and Asia, chapters reveal the ways in which exploring citizenship through an infrastructural lens, and infrastructure through a citizenship lens, allows us to better understand, plan and govern city life. The book emphasises the importance of acknowledging and understanding the dialectic relationship between infrastructure and citizenship for urban theory and practice. This book will be a useful resource for researchers and students within Urban Studies, Geography, Development Studies, Planning, Politics, Architecture and Sociology.

Charlotte Lemanski is a University Senior Lecturer in the Department of Geography at the University of Cambridge, UK. She has a particular interest in the everyday and structural realities and constraints of inequality within the Southern city, focusing specifically on inequalities related to housing and infrastructure, as well as urban governance and citizenship.

Routledge Studies in Urbanism and the City

Indigenous Rights to the City
Ethnicity and Urban Planning in Bolivia and Ecuador
Philipp Horn

Health, Wellbeing and Sustainability in the Mediterranean City
Interdisciplinary Perspectives
Edited by Antonio Jiménez-Delgado and Jaime Lloret

Tourism and Everyday Life in the Contemporary City
Edited by Thomas Frisch, Christoph Sommer, Luise Stoltenberg and Natalie Stors

Citizenship and Infrastructure
Practices and Identities of Citizens and the State
Edited by Charlotte Lemanski

Pedagogies of Urban Mobilities
Kim Kullman

Balkanization and Global Politics
Remaking Cities and Architecture
Nikolina Bobic

Cities and Dialogue
The Public Life of Knowledge
Jamie O'Brien

The Walkable City
Jennie Middleton

Urban Restructuring, Power and Capitalism in the Tourist City
Contested Terrains of Marrakesh
Khalid Madhi

For more information about this series, please visit https://www.routledge.com/Routledge-Studies-in-Urbanism-and-the-City/book-series/RSUC

Citizenship and Infrastructure

Practices and Identities of Citizens and the State

Edited by
Charlotte Lemanski

Routledge
Taylor & Francis Group
LONDON AND NEW YORK

First published 2019
by Routledge
2 Park Square, Milton Park, Abingdon, Oxon OX14 4RN

and by Routledge
52 Vanderbilt Avenue, New York, NY 10017

First issued in paperback 2020

Routledge is an imprint of the Taylor & Francis Group, an informa business

British Library Cataloguing in Publication Data
A catalogue record for this book is available from the British Library

Library of Congress Cataloging-in-Publication Data
Names: Lemanski, Charlotte, 1977- author.
Title: Citizenship and infrastructure : practices and identities of citizens and the state / Charlotte Lemanski.
Description: Abingdon, Oxon ; New York, NY : Routledge, 2019. | Includes bibliographical references and index.
Identifiers: LCCN 2018050231| ISBN 9780815385974 (hbk : alk. paper) | ISBN 9781351176156 (ebk) | ISBN 9781351176132 (epub) | ISBN 9781351176125 (mobi/kindle) | ISBN 9781351176149 (pdf)
Subjects: LCSH: Infrastructure (Economics)--Developing countries. | Citizenship--Developing countries. | Cities and towns--Developing countries. | Public welfare--Political aspects--Developing countries.
Classification: LCC HT149.5 .L45 2019 | DDC 307.7609724--dc23
LC record available at https://lccn.loc.gov/2018050231

ISBN 13: 978-0-367-66090-1 (pbk)
ISBN 13: 978-0-8153-8597-4 (hbk)

Typeset in Times New Roman
by Taylor & Francis Books

Contents

Illustrations

Figures

Tables

Contributors

Charlotte Lemanski is a Senior Lecturer in the Geography Department at the University of Cambridge. Her research broadly focuses on everyday inequality in cities of the global South, and she has explored this through various lenses; for example, housing and land markets, urban infrastructure, segregation and integration, and urban governance and citizenship. She is currently involved in inter-disciplinary research that spans the physical and social sciences, exploring domestic energy innovation for low-income urban dwellers in the global South. Her primary empirical focus is South Africa, with a secondary interest in India. She is co-editor of *The City in Urban Poverty* (with Colin Marx).

Colin McFarlane is Professor of Urban Geography at Durham University. His work focuses on the politics of urban life, including in relation to dense urban living, urban infrastructure, urban poverty, and urban inequality. His current work includes a focus on the nature and politics of urban densities in cities in Asia, urban equality in the global South, the politics of urban waste, and the material fragments of everyday life. He is author of *Learning the City: Knowledge and Translocal Assemblage* and co-editor of *Infrastructural Lives: Urban Infrastructure in Context* (with Steve Graham), *Urban Navigations: Politics, Space and the City in South Asia* (with Jonathan Anjaria), and *Urban Informalities: Reflections on the Formal and Informal* (with Michael Waibel).

Iva Peša is a historian of Central Africa, with a PhD from Leiden University (2014) on the social history of Mwinilunga, Northwest Zambia. She is currently a Research Associate in Environmental History at the University of Oxford. Her publications have explored the history of Central Africa from social, environmental and technological perspectives. Her edited books include *Transforming Innovations in Africa* (2012) and *The Objects of Life in Central Africa* (2013).

Dennis Rodgers is Research Professor in Anthropology and Sociology at the Graduate Institute of International and Development Studies, Geneva, Switzerland.

Antina von Schnitzler is Associate Professor of International Affairs and Anthropology at The New School in New York. She is the author of *Democracy's Infrastructure: Techno-Politics and Protest after Apartheid* (2016).

Jonathan Silver is a Leverhulme Early Career Fellow at the Urban Institute, University of Sheffield. He is a geographer working on global urban studies and infrastructure through comparative research across a number of cities.

Alex Wafer is a Senior Lecturer in Human Geography at the School of Geography, Archaeology and Environmental Studies, University of Witwatersrand. His most recent research project, which was funded through the National Research Foundation (2015–2017), was titled "Infrastructures of State and Citizenship". The project investigated the intersection between post-apartheid infrastructural landscapes and the forms of subjectivity and expectations of state produced by these landscapes. The project is currently being written into a book manuscript due for publication in 2019.

Foreword

Towards an infrastructural imaginary of the political

Dennis Rodgers

Infrastructure surrounds us, conditioning our lives in fundamental ways, yet we often take it for granted. In many ways, however, the history of human evolution can be conceived as a history of infrastructural development. Certainly, a widespread measure of our species' economic and political "progress" is the spread of cities from around 8,000 BC onwards, which of course could not have happened without the development of diverse infrastructure to enable unprecedentedly large, dense, and heterogeneous groups of individuals to live together within contained spaces, through, for example, the organised (re)distribution of people, goods, services, and waste. Indeed, seen from this perspective, infrastructure can arguably be said to be a precondition for social life itself. This means that infrastructure is very obviously not just a technical issue, yet despite the past few years having seen what might be termed as "infrastructural turn", there is still a predominant tendency within the social sciences to conceive of infrastructure in rather narrow, technological terms. This volume is an attempt to go beyond such perspectives and to bridge the gap to a more sociological approach, proposing the innovative theoretical lens of "infrastructural citizenship" as a means for doing so.

Like all good ideas, linking infrastructure with citizenship is obvious in hindsight, whether from a historical or a conceptual perspective. Just as infrastructure is intimately linked to the history of urbanization, so is citizenship – as the etymology of the word inherently highlights – and the former has often been a vector through which the latter is either enacted or laid claim to in cities worldwide across both time and space. Access to urban services is, for example, often a hallmark of belonging to a particular group or polity, and can therefore become an issue around which conflicts about belonging can be articulated. Conceptually, the notion of infrastructural citizenship draws attention to the fundamental connections that exist between the material and the political in the constitution of society. While not necessarily a new insight – much of 19th century social science was arguably precisely about this dialectic, albeit perhaps in a broader sense – what it allows for is a much needed opening up of the black boxes of both "infrastructure" and "citizenship" at a conjuncture when much theorising about these phenomena has perhaps become rather abstract in nature. In particular, the notion of infrastructural citizenship highlights how infrastructure is not a "neutral" phenomenon, but both shapes and is shaped by the political, and also points directly to the fact that articulations of citizenship are not abstract processes, but have very concrete material bases.

The contributions to this volume all explore different iterations of the nexus between infrastructure and citizenship, but it is striking that a common insight running through all of them is the practical primacy of the infrastructural over citizenship. At its most basic, citizenship is a fundamentally binary concept, one which posits that individuals either have or do not have certain rights associated with belonging to a particular group or polity, yet such an abstract categorisation needs to be concretely actualised in the real world. In other words, citizenship is fundamentally determined by specific modalities of inclusion and exclusion, and infrastructure is one of the key means through which these unfold, particularly in cities. The contributions to this volume highlight the diversity of ways in which infrastructure and citizenship can connect to each other, but also the way that "infrastructure" is something of a "fuzzy" phenomenon, one that can take on a range of very distinct material and non-material forms that are often fungible. Perhaps the most striking example to emerge from the contributions to this volume is the way that individual social agents can themselves become forms of infrastructure for others, particularly in contexts with deficient or only partial material infrastructural coverage. This suggests that certain types of mobilization against infrastructural non-citizenship might themselves be conceived as non-material forms of infrastructure, an almost "do-it-yourself" form of infrastructural citizenship.

This raises the obvious question of whether certain forms of infrastructure are more important or more effective than others in materially actualising citizenship. One aspect with regard to which the contributions to this volume are particularly thought-provoking is the issue of the extent to which – or not – materiality itself is critical to infrastructure's power. This is particularly important to consider because one aspect of infrastructure's connection to citizenship that is not explicitly considered in this volume, but which constitutes an inherent flipside to its central insights, is that the power of infrastructure is such that it can also become "anti-political" in nature, and be a means through which to actively deny or undermine citizenship rights by closing off the possibility for contestation via excluding the very notion of citizenship and belonging from the everyday political imaginary. Some of the protests recorded in the contributions to this volume could be read less as claims for "infrastructural citizenship" than objections to instances of "infrastructural anti-citizenship", and the obvious question to ask is to what extent such mobilizations are effective in challenging anti-political forms of "infrastructural violence".

This is important because while the contributions to this volume highlight how infrastructure can potentially be deployed and challenged in different ways, their exegeses of the notion of citizenship show it to be clearly less malleable. As such, they implicitly point to the universal normativity underpinning the concept, and more specifically to its assumed indivisibility. Numerous studies have highlighted the existence of "bifurcated", "second class", or "partial" citizenship, yet these are more often than not conceived in terms of an absence or incompleteness relative to a holistic notion of citizenship which remains the primary conceptual reference point for thinking about the political organisation of human society. In this regard, the conceptual binary of "infrastructural citizenship" highlights how there is potentially another way to conceive of the latter, one where the infrastructural constitutes a starting point, insofar as the infrastructural citizen is first and foremost infrastructural before they are a citizen. Seen from this perspective, the notion of infrastructural citizenship clearly points to fundamental issues of political economy that underpin the organisation of human society, and what this volume can be said to call for first and foremost is the spread of an "infrastructural imaginary" within our thinking about the political. Such an epistemological lens is arguably potentially much better suited to getting to grips with the dynamics of a contemporary world insofar as it is now claimed that this is characterised by processes of "planetary urbanization", and historical experience suggests that these will necessarily and fundamentally be underpinned by infrastructural processes.

Introduction

The infrastructure of citizenship

Charlotte Lemanski

This book is framed by the premise that while infrastructure has become the primary lens used within urban studies to understand the city (e.g. Graham and Marvin 2001; Graham 2010; Amin 2014; Graham and MacFarlane 2015), and citizenship is similarly promoted within political geography and development studies as vital for understanding socio-political life (e.g. Painter and Philo 1995; Isin and Nielsen 2008; Staeheli 2010; Cornwall et al. 2011; Staeheli et al. 2012), the connections between these two concepts and empirical trends are poorly understood and under-examined. Consequently, this book presents an opportunity for scholars already working on urban issues related to infrastructure and/or citizenship to reflect on the opportunities and challenges presented by adopting an infrastructural perspective to understanding citizenship (and vice versa).

I argue that connecting infrastructure and citizenship is important because it highlights the links between the material and political nature of state-society relations. While citizenship is inherently a political relationship (state-society), that relationship is frequently mediated through the materiality of public infrastructure, particularly in the urban context. Similarly, urban infrastructure is inherently a techno-material provision that is both shaped by, and has profound consequences for, socio-cultural and political life. For too long these two approaches to understanding urban life have operated within disciplinary silos that have separated not only their literatures and scholarly debates, but also limited the potential for empirical understandings of urban life to be enriched by both an infrastructure- and a citizenship-centric filter. Fundamentally, and as this book demonstrates, this is not exclusively an academic call to bridge the infrastructure and citizenship literatures, but equally an empirical call to broaden analyses of the city in order to consider both infrastructure and citizenship as fundamentally inter-connected in shaping urban life at multiple scales. Indeed, contemporary governments across the globe are simultaneously prioritising urban infrastructure as vital for economic and technological progress,[1] while at the same time promoting "citizenship" and "community" as strategies to create new forms of political identity that transcend prior identity-based (class, race, gender) politics. Consequently, public infrastructure and citizenship are at the core of the city, both in terms the state's economic-technical and socio-political urban visions, as well as the everyday lives of urban dwellers, particularly those functioning at the socio-economic and spatial margins of the urban.

It is therefore of little surprise that scholarship on both infrastructure and citizenship has become increasingly popular, and these contemporary disciplinary "turns" are represented by a vast array of workshops, conference sessions and publications focused on demonstrating and debating the utility of these broad concepts within and beyond their sub-disciplines. As a very crude summary (see Table I.1), contemporary urban scholarship conceptualises infrastructure as a technical/physical representation of socio-political processes (e.g. Larkin 2013), while political science and development studies' research on citizenship emphasises the role played by citizen action rather than legal rules per se (e.g. Cornwall et al. 2011). Collectively these two meta-narratives provide the conceptual and empirical frameworks through which scholars are currently exploring the socio-technical and legal-political ways of functioning for urban spaces, institutions, processes and people. And yet, despite the arguable connections between and parallels across their remits and approaches, there is an absence of attempts to explicitly connect the largely disparate scholarly debates that exist in relation to infrastructure and citizenship, respectively.

Table I.1, in providing a very basic summary of the core academic debates in both infrastructure and citizenship, draws out some of the most obvious connections between ideas that have separately dominated both bodies of scholarly literature, as a way of confirming both their parallel trajectories as well as the necessity of adopting a citizenship/infrastructure nexus approach. Chapter 1 of this volume, by Charlotte Lemanski, provides a more detailed critical analysis of these dual literatures. What is striking from this more detailed literature analysis is the depth of resonances, parallels and similarities in the approaches taken by scholarship on infrastructure and citizenship, respectively. For example the scale of research is typically relatively localised (the city, the settlement), human-centric (the citizen in relation to other institutions) and temporally focused on the everyday. Furthermore, in research on the global South, it is evident that urban, political and development scholarship have in parallel explored citizen-led strategies to produce and manage infrastructure, and to demonstrate citizenship rights that are typically framed around access to infrastructure. Consequently, while the connections between infrastructure and citizenship are clearly evident from even a brief reading of existing scholarship, these connections are largely implicit. For example, urban, political and development scholarship all consider the rights-based nature of access to basic urban services (e.g. Parnell and Pieterse 2010), thus implying that both infrastructure and citizenship are crucial elements for understanding the social and political urban fabric, and thus the lack of explicit debate on their relationship is surprising. Furthermore, this under-theorisation has not merely facilitated an empty void, but has allowed scholarship to implicitly define the connections between infrastructure and citizenship as primarily about radical and/or confrontational protest around access to services (e.g. Miraftab and Wills 2005; Holston 2008), often overlooking issues of everyday citizenship identity and practice, as well as the role of the state.

Table I.1 Indicative summary of core academic debates on infrastructure and citizenship

	Citizenship	*Infrastructure*	*Parallels*
Historic approach	Citizenship is the group of legal/institutional rules between the state and its citizens.	Infrastructure is a technical sector led by engineers, architects, and planners.	Simplistic conceptualisation framed by assumptions that technical-legal matters are politically/socially neutral.
Contemporary approach	Citizenship is defined by both institutional rules/laws and citizen practices.	Infrastructure is both social and technical: urban socio-political life shapes and is shaped by infrastructure.	Complex conceptualisation that acknowledges multiple rationalities and actors.
Examples of contemporary scholarship	Isin and Nielsen (2008) *Acts of Citizenship*: The processes through which citizens engage with and demonstrate their citizenship. James Holston (2008) *Insurgent Citizenship*: Citizenship can be a tool to claim rights, often in contestation with the state. Lynn Staeheli et al. (2012) *Ordinary Citizenship*: The rules and laws of citizenship are inherently tied to the everyday practices of citizens.	Susan Star (1999) *Ethnography of Infrastructure*: Infrastructure is embedded in human relations. Graham and Marvin (2001) *Splintering Urbanism:* Uneven access to infrastructure is politicised, particularly in the neo-liberal era. Graham and McFarlane (2015) *Infrastructural Lives*: Importance of the everyday ways in which people experience infrastructure (as both freedom and suppression).	Importance of human relations and actions for understanding both concepts Role of state and non-state actors in claiming infrastructure / citizenship The everyday as a core temporality in which citizenship and infrastructure function to marginalise and empower urban dwellers.

Consequently, the foundational premise of this book is that, while there is widespread recognition that both infrastructure and citizenship are crucial for understanding the everyday spaces of life in the city, the connections between them are poorly understood and under-theorised. As a means of exploring these connections this edited collection draws together scholarship that highlights the implicit and explicit ways in which infrastructure and citizenship are inherently connected. Using examples from Asia and Africa, the chapters individually and collectively demonstrate the ways in which adopting an "infrastructural citizenship" lens illuminates a broader understanding of the material and civic nature of urban life for both citizens and the state, while also demonstrating the importance of acknowledging and understanding the dialectic relationship between infrastructure and citizenship for urban theory and practice.

Lemanski's chapter, "Infrastructural citizenship: Spaces of living in Cape Town, South Africa" (Chapter 1), grounds the core arguments of this book in a critical review of the literature on infrastructure and citizenship, and demonstrates the utility of her "infrastructural citizenship" argument through a case study of housing in South Africa. Using the example of state-subsidised housing in Cape Town as a symbol for public infrastructure in the post-apartheid state, the chapter explores the connections between the physical and political elements of infrastructure and citizenship for housing beneficiaries. In this case, urban dwellers' political identity as a new citizen of South Africa is intertwined with physical receipt of public housing and associated services, whilst simultaneously the state conceives of its new citizens primarily as infrastructure consumers and/or complainers (rather than as active citizens per se). Consequently, citizenship as a political identity in the new South Africa is inherently about material identity in the sense of access to, and the quality of, public infrastructure.

Jon Silver and Colin McFarlane's chapter, "Social Infrastructure, citizenship and life on the margins in popular neighbourhoods" (Chapter 2), uses personal vignettes from what they term a "popular neighbourhood" in Kampala, Uganda. Drawing on insights from the personal stories of six diverse residents, the chapter demonstrates how social infrastructure, i.e. the ways in which people relate to and support one another in socio-material terms, enables marginalised urban dwellers to navigate the everyday precarity of urban life. Using these rich narratives, and rooted in the broader literature on marginality and precarity, the chapter makes a compelling case for social infrastructure as a lens that illuminates the ways in which marginalized urban dwellers connect with one another and with material items in order to function at the socio-economic and political margins of urban life.

Colin McFarlane's chapter, "The politics of urban sanitation: Making claims to the city" (Chapter 3), reveals the politicised nature of sanitation provision, and is framed by the two empirical case studies of Mumbai and Cape Town. Using the examples of social movements, the Right to Pee in Mumbai and the Social Justice Coalition in Cape Town, the chapter demonstrates how urban citizens claim their *right to city-life*, that is, to live in a space that represents their needs, hopes and aspirations. In both cases, citizens disrupt the state using grounded data (i.e. surveys and audits of sanitation provision, often using "smart" technologies to ensure real-time accuracy) to demonstrate the absence of state infrastructure, using the technical language of the state to demand their citizenship rights. In both cases the campaigns have shifted over time, from initial demands for specific sanitation provision and maintenance towards a broader focus on challenging inequality in the city as a whole, and the role of the state in meeting its responsibilities. In this sense, McFarlane argues that urban sanitation is inherently about the politics of the city, from the perspective of both the state and citizens.

Alex Wafer's chapter, "Enframing citizenship: Social housing and ontological orientations in Johannesburg" (Chapter 4), uses the case study of Cosmo City, a mixed-income flagship public housing project in Johannesburg, to demonstrate the ways in which South Africa's public housing functions as an example that re-orientates normative understandings of both infrastructure and citizenship. Using the example of a protest in this housing settlement, Wafer highlights the unusual nature of this protest whereby citizens did not demand greater state involvement in service provision (as in McFarlane's chapter for example), but instead protested against the state's over-involvement in enforcing building regulations (i.e. the threat of removing unlawfully constructed backyard structures). Throughout the chapter, Wafer uses the material object of the state-subsidised house as an example of a form of public infrastructure that challenges (or "ruptures") ontological understandings of infrastructure. In this sense, the house represents the infrastructure of citizenship. As citizens become recipients of public housing they are integrated into complex and messy political and economic webs of power (well beyond what was imagined by either the state or citizen), while at the same time, the state's relationship to these citizens changes through the experiences of infrastructure provision (and regulation).

Antina von Schnitzler's chapter, "Traveling technologies: Infrastructure, ethical regimes, and the materiality of politics in South Africa" (Chapter 5), provides a case study of the impact of pre-paid electricity meters on everyday urban life in South Africa. Focused on Soweto in Johannesburg, the chapter demonstrates how the installation of pre-paid meters in the post-apartheid era, with their capacity to cut off supply for non-payment, has exacerbated the already precarious lives of the urban poor. Unsurprisingly, many residents have bypassed the system and secured free access to electricity. Framed by this vignette, the chapter explores the technological politics of the ongoing battles between residents, engineers and utility officials, to re-/dis-connect meters and enforce/evade payment. While a very different form of protest from that demonstrated in earlier chapters (McFarlane, Wafer), von Schnitzler highlights the ways in which attempts to avoid technological monitoring are inherently about citizenship, particularly in the South African context where the technology itself becomes a tool of post-apartheid oppression.

Iva Peša's chapter, 'Water, housing and (in)formality in Kitwe, Zambia: Infrastructure, citizenship and urban belonging' (Chapter 6), highlights the importance of a historical perspective to understanding the connections between infrastructure and citizenship. Focused on a mining neighbourhood in Kitwe, Zambia, the chapter discusses the changing nature of infrastructure provision (water and housing) for mining employees in relation to other urban dwellers, revealing tensions based on a form of structural inequality that has its roots in the history of mining and urban governance. While historically mining employees had access to excellent infrastructure in comparison to other urban dwellers, framed through a language of legality that awarded de jure citizenship to mining employees in comparison to the de facto rights of informal dwellers, in recent years this inequality of infrastructure provision (rather than legality per se) has started to invert. Peša uses this specific case study to highlight the connections between infrastructure and citizenship for a diverse range of urban dwellers over a historic period, demonstrating the ways in which access to water and housing can not only be a consequence of citizenship status, but can also inform citizenship expectations and identities.

Note

1 For example, the "Business, Energy and Industrial Strategy" created by the UK's new Prime Minister in 2016 focuses on infrastructure to create a "Britain fit for the future" (BEIS 2017). Similarly, the new South African Prime Minister announced in 2018 that "Infrastructure … [is the] foundation for sustainable economic expansion" (Donnelly 2018).

References

Amin, A., 2014, Lively infrastructure, *Theory, Culture and Society* 31: 137–161.

Business, Energy and Industrial Strategy (BEIS), 2017, *Industrial Strategy: Building a Britain fit for the future*, White Paper Presented to Parliament by the Secretary of State for Business, Energy and Industrial Strategy by Command of Her Majesty, November 2017.

Cornwall, A., Robins, S., and von Lieres, B., 2011, States of citizenship: Contexts and cultures of public engagement and citizen action, *IDS working papers* 363: 1–32.

Donnelly, L., 2018, Infrastructure at the heart of Ramaphosa's plan to boost the economy, *Mail & Guardian*, 21 September 2018.

Graham, S. (Ed.), 2010, *Disrupted Cities: When Infrastructure Fails*, Routledge, London.

Graham, S. and Marvin, S., 2001, *Splintering Urbanism: Networked Infrastructures, technological mobilities and the urban condition*, Routledge, London.

Graham, S. and McFarlane, C., 2015, *Infrastructural Lives: Urban Infrastructure in Context*, Routledge, London.

Holston, J., 2008, *Insurgent Citizenship: Disjunctions of Democracy and Modernity in Brazil*, Princeton University Press, Princeton.

Isin, E.F. and Nielsen, G.M. (Eds.), 2008, *Acts of Citizenship*, Zed Books, London.

Larkin, B., 2013, The politics and poetics of infrastructure, *Annual Review of Anthropology* 42(1): 327–343.

Miraftab, F. and Wills, S., 2005, Insurgency and spaces of active citizenship: The story of Western Cape anti-eviction campaign in South Africa, *Journal of Planning Education and Research* 25(2): 200–217.

Painter, J. and Philo, C., 1995, Spaces of citizenship: an introduction, *Political Geography* 14(2): 107–120.

Parnell, S. and Pieterse, E., 2010, The "right to the city": Institutional imperatives of a developmental state, *International Journal of Urban and Regional Research* 34(1): 146–162.

Staeheli, L., 2010, Political geography: Where's citizenship? *Progress in Human Geography* 35(3): 393–400.

Staeheli, L.A., Ehrkamp, P., Leitner, H. and Nagel, C.R., 2012, Dreaming of the ordinary: citizenship and the complex geographies of daily life, *Progress in Human Geography* 36: 627–643.

Star, S. L., 1999, The ethnography of infrastructure, *American Behavioral Scientist* 43(3): 377–391.

1 Infrastructural citizenship

Spaces of living in Cape Town, South Africa[1]

Charlotte Lemanski

Introduction

This chapter introduces and deploys the phrase *infrastructural citizenship.* Whilst not the first to coin this phrase (see Justesen 2013; Shelton 2017), its meaning within this chapter centres on how citizens' everyday access to, and use of, public infrastructure in the city affect, and are affected by, their citizenship identity and practice. Furthermore, the phrase provides the foundations for initiating critical connections between the two scholarly fields of urban infrastructure and citizenship.

The chapter starts by reviewing existing geographical debates on citizenship, demonstrating the distinction between citizenship practices and acts, and highlighting examples of how citizenship can be framed as both "insurgent" (Holston 2008) and "ordinary" (Staeheli et al. 2012). Second, the chapter analyses scholarship on infrastructure, particularly the contemporary shift towards conceptualising infrastructure as socio-technical (e.g. Larkin 2013; Amin 2014) alongside an emerging recognition of the importance of infrastructure to the everyday lives of urban citizens (Graham and McFarlane 2015). Third, the chapter develops the notion of *infrastructural citizenship* as a conceptual lens through which to acknowledge the inter-connected relationship between infrastructure and citizenship in spaces of urban living. Finally, the chapter uses a case study of state-subsidised housing in Cape Town to critically explore how public infrastructure connects with citizenship in contemporary South Africa. This empirical case is particularly pertinent because universal citizenship is still relatively new in South Africa, and has emerged in a context where post-apartheid urban politics have been framed around infrastructural provision.

Citizenship: legal practices and socio-political acts

Citizenship has long been a central concept within work on political, development, cultural and historical geography, exploring the relationship between citizens and the state (historically the nation-state). However, contemporary scholarship has shifted focus in terms of scale and scope. Firstly, the scale of citizenship has both retracted and expanded, with recognition that citizenship functions at the local/urban scale as well as the global transnational scale, often resulting in different forms of citizenship at different scales (Marston and Mitchell 2004). Secondly, the scope of citizenship has shifted away from a static interpretation of citizenship through a legal statist framework of rights and responsibilities to a dynamic recognition of the ways in which citizens exercise and demonstrate those rights and responsibilities, and thereby in turn produce new understandings and experiences of citizenship. In this context, citizenship is not merely a legal fact, but is also an everyday act that operates at multiple scales. This shift has brought renewed energy to debates exploring the processes through which citizens engage with and demonstrate their citizenship. Engin Isin and Greg Nielsen (2008) coined the phrase "acts of citizenship", highlighting the need to move attention away from those who determine the rules of citizenship (i.e. the state) or those who act out citizenship (i.e. citizens), to instead focus on the *acts* of citizenship themselves.

> We propose to shift focus from the institution of citizenship and the citizen as individual agent to acts of citizenship – that is, collective or individual deeds that rupture social-historical patterns. (Isin and Nielsen 2008: 2)

In other words, they argue that citizenship is less about the status of individuals or institutions, and more about the processes through which citizens engage with and demonstrate their citizenship. Consequently their focus is on citizenship actions rather than on citizenship status. Isin and Nielsen (2008) further clarify their focus by distinguishing between everyday normalised and mundane *practices* of citizenship (e.g. voting, paying taxes), and *acts* of citizenship as non-repetitive and often confrontational actions taken by citizens to demand their citizenship rights. While other scholars use different labels to demonstrate citizenship acts, for example "social citizenship" (Taylor-Gooby 2008; Dwyer 2010), there is a general consensus in the literature that citizenship is not just about the legalistic contract of rights between the state and its citizens, but includes the ways in which citizens demonstrate these rights, and consequently citizenship comprises a crucial space of everyday life in the contemporary city.

At the same time, the distinctions inherent in this binary, between citizenship as a legal fact or everyday act, are challenged by the concept of *ordinary citizenship*. Coined by Staeheli et al. (2012), ordinary citizenship argues that the complexities of citizenship are best understood as part of the legal system *and* everyday activities and encounters. Consequently, rather than consider citizenship as either a legal fact or an everyday act, Staeheli et al. (2012) employ the concept of ordinariness to consider both in tandem, using the example of US immigrants, a group excluded from formal citizenship, to demonstrate their argument.

> The conceptualisation of citizenship as ordinary trains our analytical lens on the ways in which laws and social norms are entwined with the routine practices and experiences of daily lives, as citizens – and other political subjects – negotiate exclusion and marginalisation. (Staeheli et al. 2012: 629)

Ordinary citizenship emphasises the everyday and very normalised ways in which citizens demonstrate practices and acts of citizenship (in Isin and Nielsen's terms) that provide legal and social order to their daily lives. This concept is particularly useful because it acknowledges the ways in which the very ordinariness of citizenship can be an implicit mechanism for exclusion, although arguably it could also be mobilised by marginalised people as a tool for political change. This resonates with James Holston's now famous text on *Insurgent Citizenship* (2008), exploring democratic citizenship in the urban peripheries of São Paulo. Holston argues that although Brazilian citizenship is universally inclusive in legal terms, it is inegalitarian in practice based on unequal access to land and services (i.e. infrastructure). Holston distinguishes between what he terms modernist citizenship, whereby citizenship is restricted to those who submit to state-led expectations and norms, and his notion of insurgent citizenship, whereby citizens challenge the state, often in violent and confrontational ways, for example illegally building houses and invading land as a means to claim their citizenship rights. This approach is pertinent because it highlights the ways in which (modernist) citizenship can be exclusionary, and how (insurgent) citizenship can empower marginalised communities to take action against the state in order to secure a city that better reflects their interests and needs.

To conclude this section, within the contemporary focus on citizenship as incorporating both legal practices and socio-political acts, there is widespread recognition that citizenship plays a role in both marginalising certain people (e.g. via legal frameworks and ordinary everyday acts) as well as being a tool for marginalised groups to protest (e.g. via insurgency or more "ordinary" strategies) at multiple scales. Within this context there is an implicit assumption that those perceived as "marginalised" are those with limited access to material goods and public infrastructure (often there are multiple marginalisations, e.g. immigrants lacking legal status, but weak access to infrastructure is typically connected), and yet the explicit connections between citizenship and infrastructure are rarely considered in depth. The next section provides an overview of contemporary scholarship on urban infrastructure, before the chapter turns to consider the concept of *infrastructural citizenship* as a means to explicitly connect these scholarly debates.

Infrastructure: socio-technical space of everyday life

While infrastructure has always been an implicit component of both the urban and development geography sub-disciplines, it has received a surge of interest from contemporary social scientists, recognising that infrastructure is not a neutral technical matter to be addressed only by engineers, planners and architects, but vital to social life. Within this contemporary perspective, infrastructure is conceptualised as more than just the physical manifestation of the city or the material means through which the urban is able to function. For example, cities and urban dwellers certainly need roads, electricity, water, sanitation and housing, which are all provided through the technology of road, pipes, cables and wires; but infrastructure is also inherently social in the way it is both produced and used. Within this context there is widespread recognition in the literature that a critical understanding of urban infrastructure requires a multi-disciplinary approach, spanning the physical and social sciences, to acknowledge the ways in which infrastructure represents a physical manifestation of a social *and* technological process, in terms of both construction and consumption.

Tracing the emergence of these socio-technical debates, Susan Star's (1999) *Ethnography of Infrastructure* broke new ground by challenging existing conceptualisations of infrastructure as exclusively technical. Through her ethnographic fieldwork working alongside biologists, medics and computer scientists, she revealed the ways in which infrastructure design is embedded in humans and their relationships. Within geography, Stephen Graham and Simon Marvin's now landmark *Splintering Urbanism* (2001) text has been at the forefront of new social science approaches to infrastructure. Graham and Marvin (2001) frame urban infrastructure as a socio-technical process, highlighting the ways in which the spaces of urban societal life (e.g. shopping, cooking, washing, sanitation) are made possible by technology, and consequently they reveal how the human body and technological machines are mutually dependent. Building on this approach, urban and cultural geographer Matthew Gandy (2004) uses the human biology concept of metabolism to unite the social and biophysical systems that underpin the city; while social anthropologist AbdouMaliq Simone's (2004) conceptualisation of *people as infrastructure* situates urban infrastructure not in roads and pipes, but in the social and economic collaborations that form between marginalised urban citizens.

Alongside the now widespread recognition that infrastructure is not merely technological, but also crucial to understanding contemporary urban societal life, comes increasing concern over unequal access to infrastructure. Graham and Marvin (2001) coined the phrase 'splintering urbanism' to describe the exclusionary and fragmentary effects of privatising infrastructure provision. In the neoliberal era of privatisation, the infrastructural division of the city is driven by market rationalities, but has clear social implications; where water pipes bypass poor communities, or new train lines connect wealthy residential and commercial zones (but ignore poorer parts of the city), or where internet cables are prioritised in wealthy neighbourhoods. Consequently, urban splintering is both spatial and social, dividing the city into well-connected and under-connected zones, with societal consequences for poverty and inequality. While historically infrastructure was viewed as neutral or technocratic, Graham and Marvin's (2001) approach reveals how networked infrastructure is a tool of social power that can extend and perpetuate inequality.

Furthermore, this inequality is not merely about access to physical goods and services, but reveals broader processes of exclusion and marginalisation that are intrinsically connected to citizens' rights and the politics of the city.

Most recently there has emerged a shift towards exploring infrastructure from the perspective of citizens' everyday social and political lives. Graham and McFarlane's (2015) *Infrastructural Lives* focuses on the everyday scale as the means to illuminate citizens' perceptions and experiences of infrastructure.

> While infrastructure debates have made important contributions to how we understand the "supply-side" dimensions of infrastructure, there has been surprisingly little about how people produce, live with, contest, and are subjugated to or facilitated by infrastructure. (Graham and McFarlane 2015: 2)

The central goal of this everyday perspective is to explore how urban dwellers produce and manage their own infrastructure needs, in order to better understand the connections between urban life and infrastructure. Essentially, this is about exploring the relationship between infrastructure and everyday urbanism across a variety of empirical contexts, and consequently this perspective explicitly acknowledges infrastructure as a space of urban life. Similarly, urban geographer Ash Amin (2014) has analysed the sociality of infrastructure from the perspective of the poor, using a Brazilian case study to highlight the ways in which the poor themselves play a role in co-constructing infrastructure (e.g. water, electricity, shelter, sanitation), consequently highlighting how infrastructure is a crucial component of both individual and collective spaces of living and protest. This resonates with Simone's notion of *relational infrastructure* (2015), exploring low-income urban dwellers' incremental "pay as you go" strategies to improve and repair infrastructure as demonstrating people-led (rather than state-led) infrastructure. There seems a clear connection here, between three recent bodies of scholarship on infrastructure – Graham and McFarlane's (2015) *infrastructural lives*, Amin's (2014) *lively infrastructure* and Simone's (2015) *relational infrastructure* – and Holston's (2008) classic text on *insurgent citizenship.* All four explicitly recognise poor people's self-led strategies to provide their own infrastructural needs in the absence of adequate state provision, yet only Holston explicitly frames people-led infrastructure as a demonstration of urban citizenship. While Graham and McFarlane (2015) focus their everyday lens through questions related to violence, pacification and dispossession, with a special interest in waste, they acknowledge that this perspective is incomplete. Arguably, understanding infrastructure from the perspective of urban dwellers themselves has a clear connection to citizenship. While the physicality of inequality manifests in access to infrastructure, this is a material representation of unequal citizenship in the dual sense of both urban dwellers' access to citizenship rights as well as citizens' differentiated mobilisation capacities.

> Infrastructure is proposed as a gathering force and political intermediary of considerable significance in shaping the rights of the poor to the city and their capacity to claim those rights. (Amin 2014: 1)

The increasingly widespread conceptualisation of infrastructure as both a technical and social element of institutional and everyday urban life has resulted in recognition of the role of infrastructure in other areas of urban life, such as urban violence (Rodgers and O'Neill 2012) and urban politics (McFarlane and Rutherford 2008), highlighting how broader processes of marginalisation are operationalised through infrastructure in contemporary cities. This chapter argues the need for a more explicit connection between infrastructure and citizenship, using the phrase "infrastructural citizenship" as the means for improved understanding of both concepts and practices, and the relationship between them.

Infrastructural citizenship

The connection between infrastructure and citizenship seems obvious, and indeed is often implicitly acknowledged in the literature. For example, there is widespread recognition that those with restricted citizenship rights (e.g. immigrants, homeless, slum-dwellers) suffer multiple marginalisations which typically include weak access to infrastructure such as water, housing and sanitation (e.g. Roy 2003; Staeheli et al. 2012). Similarly, the concept of *propertied citizenship* promotes an ideal of "proper" citizenship, restricted to those whose legally entrenched access to infrastructure (via property ownership) gives them a vested stake in the nation-state that is physical, economic, political and social (Roy 2003). Indeed, Partha Chatterjee (2004) highlights the ways in which those with secure access to infrastructure (e.g. the middle class) use this advantage to stress their superior claim to citizenship (as law-abiding tax-payers) in order to further marginalise those with weak infrastructural status in the city. Furthermore, contemporary approaches to infrastructure as socio-technical highlight the ways in which infrastructure provides a physical representation of broader socio-political processes that implicitly include citizenship practices (e.g. "rights of the poor to the city", Amin 2014: 1) and acts (e.g. "capacity to claim those rights", Amin 2014: 1). And as a final example, the literature is rife with examples of citizenship acts that are not only focused on demanding improved access to infrastructure but frequently use infrastructure as a tool of protest (e.g. road blocks, gaining access to services illegally, self-built homes), such as Holston's insurgent citizenship (2008). However, despite these clear examples of the connections between infrastructure and citizenship, the relationship is rarely explicitly acknowledged or critically analysed, and is certainly under-theorised. In this chapter I propose the phrase *infrastructural citizenship* as a conceptual lens to tackle these oversights. Its definition is necessarily broad, focusing attention on the ways in which citizenship acts and practices are embodied in public infrastructure (and vice versa), in order to deepen understanding of the infrastructure–citizenship nexus in both theoretical and empirical terms. This is important because it explores potential connections between the infrastructural and civic nature of state-citizen relations.

This chapter is not the first to coin the phrase. Rune Justesen (2013), in a blog essay entitled "Infrastructural citizenship", uses the concept to expose the active role of infrastructure in citizen-led forms of political mobilisation. Framed around a case study of the Los Angeles port closures, Justesen's (2013) argument that humans are not merely replaceable parts of the port system but are completely integral to infrastructure, resonates with both Star's (1999) revelation that infrastructure design is embedded in human relations, as well as Simone's emphasis on *people as infrastructure* (2004). For Justesen, the inter-dependency between infrastructure and humans is confirmed by the closure of the port following strikes in 2012. Consequently, Justesen argues that infrastructure such as the port functions as a political space where active citizenship is demonstrated, for example via protests, strikes and occupations; and where citizens are fundamental to the effectiveness of infrastructure. This resonates with Rosalind Fredericks' (2018) *Garbage Citizenship* argument that the public dumping of household waste in Dakar demonstrates the emergence of a form of citizenship in which infrastructure functions as a tool for political action. While these examples highlight the overlaps and connections between radical demonstrations of citizenship and sites of infrastructure, the emphasis on the physical site (e.g. the port and its workers for Justesen), or the service (e.g. the garbage-based forms of protest for Fredericks), as the vessel for infrastructural citizenship is too narrow. As this chapter demonstrates, infrastructural citizenship is not restricted to those physically and economically positioned within infrastructure sites, or those engaged in radical protest, but instead has relevance for all citizens who (sometimes struggle to) access infrastructure in both everyday *and* radical ways.

Although not explicitly deploying the term infrastructural citizenship, Alex Wafer's (2012, and this volume) research in South Africa on discourses of infrastructure and citizenship critically explores the relationship between the state's infrastructural capacity and imaginations of citizenship in everyday life within contemporary Johannesburg. Wafer (2012) dramatically highlights the centrality of infrastructure to citizenship in South Africa: from the ways in which infrastructure represented the power of the apartheid state, for example via urban segregation and the forced removal of non-citizens; to the post-apartheid government's agenda of service delivery as the mechanism for operationalising universal citizenship; to the contemporary context of violent protests related to service delivery and illegal occupation of urban infrastructure as active demonstrations of citizenship. Wafer's (2012) Soweto case study reveals the ways in which the inherent connection between infrastructure and citizenship has changed over time, as he traces his involvement in the community from analysing protest-based citizenship (infrastructural in the sense that protests are framed around access to services) in the early 2000s, to witnessing more consumption-based citizenship in the 2010s (infrastructural in being framed around new shopping malls and materialism). Wafer's (2012) analysis is crucial in highlighting the embeddedness of the infrastructure–citizenship nexus in South Africa.

Most recently, Kyle Shelton's (2017) historical analysis of Houston residents' strategies of mobilisation against highway construction in the 1970s uses infrastructural citizenship to highlight the ways in which the inert and technical materiality of infrastructure has political meaning. Shelton's rich historical analysis uses this case study to demonstrate how residents used infrastructure as a tool for political protest by arguing that their local streets and homes held equal value as major highways and urban regeneration projects, consequently using infrastructure as the means to assert their citizenship rights. In this context, the physicality of infrastructure took on a political and symbolic dimension as the location of citizenship practices and acts. Shelton uses this example to argue that citizen protests over infrastructure must not be sidelined as a minority or radical concern, but instead that infrastructure (both in a physical and symbolic sense) should be understood as central to urban citizenship. While embracing Shelton's call, and echoing the need to recognise the connections between infrastructure and citizenship, in this chapter I extend the concept beyond the protest-based forms of infrastructural citizenship that Shelton (2017), Fredericks (2018), Wafer (2012) and Justesen (2013) all describe, to also consider forms of infrastructural citizenship that are not framed around radical or confrontational demonstrations, but are integral to everyday spaces of urban life.

This focus on the role of infrastructure in the "everyday" rather than the "spectacular" temporality and spatiality of the city resonates with Nikhil Anand's (2017) *Hydraulic City*, where he identifies the ways in which practices of citizenship in Mumbai are illuminated by an exploration of water infrastructure in the city. Through an in-depth study of water access he identifies the ways in which access to water is not simply secured via pipes, but grounded in human relations (e.g. repair engineers) and rooted in unequal practices that reflect and reproduce existing inequalities in the city. Furthermore, he demonstrates the ways in which access to formal water infrastructure becomes a leverage for access to other forms of public infrastructure, thereby revealing the ways in which infrastructure operates as a conduit for citizenship.

Having introduced the concept of infrastructural citizenship, the chapter proceeds by exploring its utility within the empirical context of contemporary South Africa.

Infrastructural citizenship in Cape Town, South Africa

Within scholarship addressing contemporary urbanisation in sub-Saharan Africa, there has been a marked focus on urban infrastructure, particularly in the context of rapid urbanisation, hyper-fragmentation and low household incomes. African urbanist Edgar Pieterse (2014) argues that the expansion of slum urbanism has resulted in a polycrisis of extreme splintered urbanism with significant inequalities in both access to, and the quality of, infrastructure. More specifically, Pieterse and Hyman (2014) express concern regarding the extreme inequality represented in the current African model where the wealthy rely on private sector provision and the poor rely on self-construction. The theoretical implications of this urban political ecology have been situated within the African context by Lawhon, Ernstson and Silver (2014), providing critical insights into the sociality and materiality of urban infrastructure in Africa, with a particular focus on the unequal distribution of resources. This city-scale approach is complemented by localised studies exploring the specific hurdles faced particularly by low-income urban dwellers in accessing basic services and infrastructure such as electricity, water and sanitation, revealing the ways in which poor Africans circumvent state infrastructure in order to independently construct or secure access (e.g. Jaglin 2014; Robins 2014; Baptista 2015). This emphasis on citizen-led infrastructure provision echoes trends identified elsewhere (e.g. Amin 2014; Graham and McFarlane 2015; Simone 2015).

Concurrently, scholarship on citizenship in urban Africa has received renewed attention, particularly in the context of democratisation. Dorman, Hammett and Nugent (2007) describe the ways in which citizenship has been re-framed in contemporary Africa around new forms of statehood and national belonging, in contrast to historic ethnic forms of belonging. They argue that state-based citizenship provides a framework for state–society relations. Building on this, Diouf and Fredericks (2014) define their *arts of citizenship* as the innovative and everyday ways in which African urban dwellers negotiate their citizenship rights and rewards, for example via involvement in churches, community groups and political organisations. They explore these arts in the context of infrastructures as a physical representation of the state and of broader processes of inclusion and exclusion, revealing "how delinations of citizenship are codified and contested in the built form of the city" (Diouf and Fredericks 2014: 6). In this sense, there is a clear connection between citizen-constructed infrastructure (e.g. house-building, electricity connections) and citizen-negotiated citizenship practices (e.g. associational life), particularly for low-income citizens. This is pertinent in revealing more about state–society relations in the context of infrastructural citizenship. Indeed, it seems obvious that infrastructure is not only one of the primary mediums through which citizens relate to the state (in terms of both expectations and physical realities), but also the primary way in which the state is physically represented at the local scale, and can play a role in creating and perpetuating (or potentially challenging and overcoming) marginalisation.

Shifting attention specifically to South Africa, this provides a particularly interesting case study for exploring infrastructural citizenship because universal citizenship is only two decades old and has been accompanied by a raft of infrastructural-based policies. The post-apartheid government has prioritised an infrastructure-centric vision of citizenship, with the provision of housing and associated services (e.g. water, sanitation, electricity) implemented as the state's primary mechanism for overcoming past inequalities and implementing citizenship (Swilling 2006; von Schnitzler 2008; Parnell and Pieterse 2010; Wafer 2012). Apartheid South Africa provided a classic historic example of excluded citizenship. During apartheid, citizenship was exclusively reserved for the white population, with other races designated citizens of their rural "homeland" and therefore ineligible to own property, vote, or move freely in South Africa's cities. This was clearly and explicitly not universal citizenship, and was inherently connected to infrastructure in terms of differentiated access. Furthermore, the collapse of apartheid in the late-1980s was in part a consequence of infrastructure-based protest, as black urbanisation gradually threatened the apartheid system of urban segregation. In contrast, political change came more abruptly. On 27 April 1994, for the first time, all South Africans were legally considered full citizens with the right to vote in their new democracy. The new constitution and bill of rights came into force, with a particular emphasis on freedom and equality for all citizens. However, the spatial, social and economic legacies of apartheid remain, and in the contemporary era, mass protests over access to services (e.g. housing, electricity, water, jobs, education) have come to represent citizenship struggles for the urban poor. For while South Africa's progressive bill of rights gives full citizenship rights to all, access to services is still skewed according to income and spatial location, and thus access to infrastructure continues to play a major role in mediating perceptions and practices of citizenship.

The most significant infrastructural provision in the post-apartheid context is the construction of housing for low-income households. Since 1994 more than three million state-subsidised fully serviced brick-built houses have been constructed and occupied (alongside full ownership rights). Despite criticisms of the housing policy in terms of the quality of construction, size of houses, location of settlements, delays in receiving title deeds, and tendency to view beneficiaries as static recipients rather than active citizens, the scale of construction is impressive. For many housing beneficiaries there is significant symbolic association with becoming a citizen and receiving ownership of a fully serviced state-subsidised house located in a city in which they were previously considered a temporary sojourner. Based on fieldwork undertaken over a long period (2004–ongoing) in a single state-subsidised housing settlement in Cape Town, South Africa, I use residents' perceptions to indicate the potential utility of infrastructural citizenship as a lens through which to understand these experiences.

What is clear from over a decade of fieldwork in this state-subsidised housing settlement is that housing beneficiaries themselves demonstrate a strong connection between citizenship and infrastructure. In particular, residents frequently expressed a vivid association between the home that they now own and "the end of the struggle". In this sense, housing beneficiaries demonstrate explicit attachment to the physicality of the house and its associated infrastructure in terms of their political identity as citizens of a post-apartheid South Africa and of Cape Town, a country and a city where they have potentially always lived, but have only recently received citizenship.

> Do you know how much this house is worth to me? Everything. I cannot give a price because this is the first and last house I have ever owned. [A.D. 03/09/06][2]
>
> If I ever get the opportunity to buy somewhere bigger, but I will never sell that house. I would rent it out because that house means so much to me, that's where all my dreams are. The first home I built for my family. It's a house that gave me strength … I won't just give it away, it's very close to my heart. [M.N. 14/09/06]
>
> If a person asked me to sell my house they can go high and high, it doesn't matter. I won't sell it because I stay here … we struggled a lot and the Lord blessed me with this house. [Ry.M. 03/09/06]

As these quotations demonstrate, the start of a new life in a democratic and free nation is emotionally and symbolically tied not merely to homeownership *per se*, but to the specific physical house and infrastructural services that embody this freedom. However, there are also dissenting views, with numerous residents expressing a more negative association with the physicality of their house and services.

> I was very unhappy when I moved in because we struggled for years and we got kak [shit]. You see the cracks in the wall, there's a lot, in the kitchen, bathroom and living room … They don't care about us here. [C.M. 16/09/06]
>
> All the houses is falling apart – the doors, the windows, it all fall apart. They build it too quickly – there's a lot of cracks. I think they built in a month. They just wanted us to move out so the rich people can move in. They promised us a lot and none of it came. [L.T. 14/09/06]

Although these latter quotations demonstrate less contentment with the material properties of the house, the relationship between the physical embodiment of a state-subsidised house and the sense of citizenship in the new South Africa remains prominent. While residents' perception of the quality of their house differed (although in fact, even those who highly valued their house still recognised its physical failings), the material provision of public housing and infrastructure is clearly framed around citizenship. While for some receiving a state-subsidised house embodies a sense of equal citizenship in post-apartheid South Africa, for others it highlights perceptions of the perpetuation of inequality and marginalisation in the post-apartheid era that is now framed by the language of citizenship, particularly in comparison to the material wealth of other citizens.

In situating these empirical findings within contemporary literature exploring the connections between infrastructure and citizenship, a disjuncture emerges. While Holston (2008) and Amin (2014) both demonstrate how marginalised citizens construct their own forms of infrastructure, in this South African case study, the state has provided physical infrastructure in the form of housing and associated services. Consequently infrastructural citizenship emerges in the ways in which citizens relate to this state-provided infrastructure. Returning to Isin and Nielsen's (2008) distinction between citizenship practices and acts, this South African example demonstrates citizenship identity as rooted in the legality of access to infrastructure in the post-apartheid context of universal citizenship, with housing beneficiaries sufficiently confident in their new citizenship rights to criticise the quality and inequality of public infrastructure. This citizenship is rooted in legal *practices*, but demonstrated through public infrastructure as the physical representation of the state in citizens' everyday spaces of living. Although these quotations do not reveal the extent of citizenship *acts*, in terms of how citizens actively demonstrate their infrastructural citizenship, there is widespread evidence for example that housing beneficiaries demonstrate insurgent citizenship (Holston 2008) acts via building (illegal) informal structures in their backyards and accessing electricity extensions (see Lemanski 2009). Furthermore, it is evident that associational life as a form of arts (Diouf and Fredericks 2014) and/or ordinary (Staeheli et al. 2012) citizenship acts has been facilitated via the provision of social infrastructure in terms of churches and sports groups, as well as providing the political focal point for community groups (see Lemanski 2006, 2008). Arguably then, this case study reveals how infrastructure both shapes citizenship practices and identities as well as determining citizens' capacity to engage in, as well as providing a focal point for, citizenship acts.

Conclusion

This chapter has introduced the concept of infrastructural citizenship as a conceptual lens through which to connect scholarly debates on infrastructure and citizenship. This approach resonates with research situated elsewhere recognising the ways in which citizens demonstrate their citizenship practices and acts (Isin and Nielsen 2008) via the self-construction of infrastructure (e.g. Holston 2008; Amin 2014; Graham and McFarlane 2015; Simone 2015). Using a case study of state-subsidised housing in South Africa, the chapter has revealed the ways in which the legalities of citizenship practice, perceptions of citizenship identity, and expressions of citizenships acts are all embedded in the physicality of public infrastructure as a representation of the state at the local scale. In this specific case study, housing beneficiaries express citizenship as embodied by their housing, with physical receipt of a house representing the start of a new identity as a South African citizen, accompanied by the confidence to criticise the state via infrastructure. This "ordinary" approach to citizenship (Staeheli et al. 2012) allows the inclusion of a broad range of citizenship meanings, from everyday associational life to protest-based conceptualisation of citizenship, in all cases highlighting the role of public infrastructure as the physical means through which citizens demonstrate their citizenship identity, practice and acts.

Notes

1 Some of the material in this chapter previously appeared in *The Routledge Handbook on Spaces of Urban Politics*, and is reproduced here with the publisher's permission.
2 Quotes from interviewees are identified by anonymous initials and the date of the interview.

References

Amin, A., 2014, Lively infrastructure, *Theory, Culture and Society* 31: 137–161

Anand, N., 2017, *Hydraulic City: Water and the Infrastructures of Citizenship in Mumbai*, Duke University Press, Durham, NC.

Baptista, I., 2015, "We live on estimates": Everyday practices of prepaid electricity and the urban condition in Maputo, Mozambique, *International Journal of Urban and Regional Research* 39(5): 1004–1019.

Chatterjee, P., 2004, *The Politics of the Governed: Reflections on popular politics in most of the world*, Columbia University Press, New York.

Diouf, M. and Fredericks, R. (Eds.), 2014, *The Arts of Citizenship in African Cities: Infrastructures and spaces of belonging*, Palgrave Macmillan, New York.

Dorman, S., Hammett, D. and Nugent, P. (Eds.), 2007, *Making Nations, Creating Strangers: States and citizenship in Africa*, Koninklijke Brill, Leiden.

Dwyer, P., 2010, *Understanding Social Citizenship: Themes and perspectives for policy and practice*, Policy Press, Bristol.

Fredericks, R., 2018, *Garbage Citizenship: Vital infrastructures of labor in Dakar, Senegal*, Duke University Press, Durham, NC.

Gandy, M., 2004, Rethinking urban metabolism: Water, space and the modern city, *City* 8: 363–379.

Graham, S. and Marvin, S., 2001, *Splintering Urbanism: Networked infrastructures, technological mobilites and the urban condition*, Routledge, London.

Graham, S. and McFarlane, C., 2015, *Infrastructural Lives: Urban infrastructure in context*, Routledge, London.

Holston, J., 2008, *Insurgent Citizenship: Disjunctions of democracy and modernity in Brazil*, Princeton University Press, Princeton.

Isin, E.F. and Nielsen, G.M. (Eds.), 2008, *Acts of Citizenship*, Zed Books, London.

Jaglin, S., 2014, Regulating service delivery in southern cities: Rethinking urban heterogeneity. In S. Parnell and S. Oldfield (Eds.) *The Routledge Handbook on Cities of the Global South*, Routledge, London: 434–447.

Justesen, R., 2013, *Infrastructural Citizenship*, https://infrastructuralgossip.wordpress.com/2013/09/15/infrastructural-citizenship/(accessed 21/01/2016).

Larkin, B., 2013, The politics and poetics of infrastructure, *Annual Review of Anthropology* 42(1): 327–343.

Lawhon, M., Ernstson, H. and Silver, J., 2014, Provincializing urban political ecology: Towards a situated UPE through African urbanism, *Antipode* 46(2): 497–516.

Lemanski, C., 2006, The impact of residential desegregation on social integration: Evidence from a South African neighbourhood, *Geoforum* 37: 417–435.

Lemanski, C., 2008, Houses without community: Problems of community (in)capacity in a low-cost housing community in Cape Town, South Africa, *Environment and Urbanization*, 20(2): 393–410.

Lemanski, C., 2009, Augmented informality: South Africa's backyard dwellings as a by-product of formal housing policies, *Habitat International* 33(4): 472–484.

Marston, S.A. and Mitchell, K., 2004, Citizens and the state: Citizenship formations in space and time. In C. Barnett and M. Low (Eds.), *Spaces of Democracy: Geographical Perspectives on Citizenship, Participation and Representation*, Sage, London: 93–112.

McFarlane, C. and Rutherford, J., 2008, Political infrastructures: Governing and experiencing the fabric of the city, *International Journal of Urban and Regional Research* 32(2): 363–374.

Parnell, S., and Pieterse, E., 2010, The "right to the city": Institutional imperatives of a developmental state, *International Journal of Urban and Regional Research* 34(1): 146–162.

Pieterse, E., 2014, How can we transcend slum urbanism in Africa? UN-Habitat Global Urban Lecture Series. http://unhabitat.org/how-can-we-transcend-slum-urbanism-in-africa-edgar-pieterse-university-of-cape-town/ (accessed 21/01/2016).

Pieterse, E. and Hyman, K., 2014, Disjunctures between urban infrastructure, finance and affordability. In S. Parnell and S. Oldfield (Eds.) *The Routledge Handbook on Cities of the Global South*, Routledge, London.

Robins, S., 2014, Poo wars as matter out of place: "Toilets for Africa" in Cape Town, *Anthropology Today* 30(1): 1–3.

Rodgers, D. and O'Neill, B., 2012, Infrastructural violence: Introduction to the special issue, *Ethnography* 13(4): 401–412.

Roy, A., 2003, Paradigms of propertied citizenship: Transnational techniques of analysis, *Urban Affairs Review* 38(4): 463–491.

Shelton, K., 2017, Building a better Houston: Highways, neighbourhoods, and infrastructural citizenship in the 1970s, *Journal of Urban History* 43(3): 421–444.

Simone, A., 2004, People as infrastructure: Intersecting fragments in Johannesburg, *Public Culture* 16(3): 407–429.

Simone, A., 2015, Relational infrastructures in postcolonial urban worlds. In S. Graham and C. McFarlane (Eds.), *Infrastructural Lives: Urban infrastructure in context*, Routledge, London: 17–38.

Staeheli, L.A., Ehrkamp, P., Leitner, H. and Nagel, C.R., 2012, Dreaming of the ordinary: Citizenship and the complex geographies of daily life, *Progress in Human Geography* 36: 627–643.

Star, S.L., 1999, The ethnography of infrastructure, *American Behavioral Scientist* 43(3): 377–391.

Swilling, M., 2006, Sustainability and infrastructure planning in South Africa: A Cape Town case study, *Environment and Urbanisation* 18(23): 23–50.

Taylor-Gooby, P., 2008, *Reframing Social Citizenship*, Oxford University Press, Oxford.

von Schnitzler, A., 2008, Citizenship pre-paid: Water, calculability and techno-politics in South Africa, *Journal of Southern African Studies* 34(4): 899–917.

Wafer, A., 2012, Discourses of infrastructure and citizenship in post-apartheid Soweto, *Urban Forum* 23: 233–243.

2 Social infrastructure, citizenship and life on the margins in popular neighbourhoods

Jonathan Silver and Colin McFarlane

Introduction

How do residents learn to navigate the topographies of everyday life in the city and on the margins of citizenship? An old question for research on cities, development and poverty, to be sure, but one that now has a new urgency and is the focus of this chapter. In particular, the growth of informality – i.e. precarious housing, labour and lives predominantly in the global South, but increasingly in the global North too (Comaroff and Comaroff 2015) – demands that we develop a better understanding of how people live in the city. Our concern is with how people cope, work together, deal with threats and develop opportunities, and invest their energies in the making of urban life, often in conditions of poverty and marginality in which the value of citizenship is barely visible (Rao 2010). To do so, we investigate the relations between people and space in and beyond one "informal settlement" – we prefer the term "popular neighbourhoods", as we set out below – in Kampala, Uganda. We seek to contribute to debates on urbanisation by demonstrating that a focus on how residents navigate the city can extend our understanding of the experience of everyday life, and by examining the role and limits of "social infrastructure" for coping with inequality and poverty. The term social infrastructure is meant to highlight the practice of connecting people and things in socio-material relations that sustain social reproduction on the margins of economic life and citizenship.

Our opening question above has been with urban research in one form or another throughout its history. Whether Friedrich Engels (1987) on Manchester or Jane Jacobs (1961) on New York or AbdouMaliq Simone (2014) on Jakarta, or in the landmark writings of Walter Benjamin (2003) and Michel De Certeau (2011[1984]) – to name just a few – the nature, relations and potential of the everyday life of the city has been interrogated from an array of different perspectives. As Henri Lefebvre (1991[1958]) argued, this is the time-space through which life takes place, through which multiple fragments coalesce into routines, practices and social events. There is a growing literature focused largely on the global South that examines the making of everyday life in marginal urban contexts (see, for example, Banks 2016; Bayat 2010; De Boeck 2015; Gill 2000; Kihato 2011; Lomnitz 1977; Moser 1996, 2010; Silver 2014; Simone 2010,

2014). These debates are increasingly being drawn on to understand everyday urban life in global North contexts, especially in the context of growing forms of precarious urban living (Hall 2012, 2015; Lancione 2014; Roy and Crane 2015; Vasudevan 2015).

Our context is everyday life at the margins of global urbanism. "Margins" of course, are produced and experienced in different ways in different cities in relation to citizenship. As Janice Perlman (2006) has argued, the term marginality is sometimes used interchangeably with "social exclusion", at other times in connection to curtailed citizenship rights (e.g. Holston 1999), or even "capabilities deprivation" (following Amartya Sen). Others use the term "advanced marginality" (Wacquant 2008) to denote the combination of neoliberal exclusion and punitive criminalisation in stigmatised urban places in which the rights of so-called citizenship are restricted. While Perlman had argued in 1979 in *The Myth of Marginality* that marginality was a "myth" in that favelas in Rio were not separate from society, but fully integrated, even if they were economically exploited, socially stigmatised and citizens only in name. By 2006 she used the term to describe the changing forms of exploitation in favelas, from growing violence and inequality to stigmatisation and a reduced sense of community life. In *Favela* (Perlman 2010), she distinguishes between material, cultural, historical, socio-psychological and political dimensions of marginality, which she takes to be mutual, reinforcing and differentiated across favelas. What this underlines is the need to understand the marginality of citizenship as a multi-dimensional struggle of but also beyond poverty, a struggle to get by and get on in cities that too often materially and discursively place limits on urban lives through capitalist and social inequalities. Here, Perlman's analysis resonates with the work of Holston (2008) and Calderia (2000) and their discussions on "disjunctive citizenship", where legal citizens are excluded from the practices of citizenship that extend to the right to access various infrastructural services. Calderia (2000) takes this notion further in her book *City of Walls* distinguishing between political citizenship, with the emphasis on "legal citizens", and civil citizenship, which she argues is the experience of life in which the failure of the state to provide service provision is central to the everyday. As Brown et al. (2010: 668) argue, "In practice, citizenship is sometimes denied to the poor and marginal by powerful elites who assign to them 'outsider status', often used as justification for violent expulsions."

The popular neighbourhood is an economic and very often citizenship margin, but a social centre of life in the urban 21st century, and is radically differentiated across the urban world economically, politically, socially and culturally. The term "popular" has a double meaning: an emphasis on populous and a play on the notion of choice – a forced choice, of course – to locate in a particular place, given that other more expensive areas of the city are often off-limits. The term "neighbourhood" rather than "settlement" serves to remind us that informal settlements are more than just sites and shelter; they are actual urban places with lives and goings-on, much like other neighbourhoods (e.g. see Hansen and Verkaaik 2009).

As the world becomes ever more urban, and as more and more of that urban population lives in some form of popular neighbourhood, it will be increasingly vital that we develop conceptual vocabularies that help us to make sense of how everyday life is made. In our previous work in Indian, South African, Ghanaian and Ugandan cities, we have examined how people learn to live in cities under conditions of precariousness, uncertainty and without the basic rights of citizenship (McFarlane 2011; McFarlane and Silver 2017; Silver 2014; Zeiderman et al. 2015). In this chapter, we argue that everyday urbanism is produced through an ongoing, co-evolving movement of contradiction, reinforcement, fragmentation and reconstitution in relation to the limits of citizenship and state support.

These dynamic relations give rise to key processes through which urban life proceeds, but they take shape differently for different people, as we will see. In the relative absence of key provisions from the state – basic welfare, social services and infrastructures – residents cope with the challenges of life in the city, and seek to improve their conditions, through the development of infrastructures and practices of different sorts (Simone 2004; Silver 2015). In particular, we identify four: social infrastructure, coordination, consolidation and speculation, all of which are experienced to a greater or lesser extent, in distinct ways and through different geographies, by different people as they are produced dialectically in the shifting assemblages of urban life.

The context of Kampala, Uganda

Kampala is the capital and largest city of Uganda, located on the northern shores of Lake Victoria. A city with an annual growth rate of 4 per cent and over 2.1 million residents (UN-Habitat 2011), Kampala is not a megacity, nor is it particularly prominent in the trajectories of global urban policy or research debate (e.g. Roy and Ong 2011). Nonetheless, like many cities in the global South it has ambitious plans (Watson 2014). Redevelopment plans connect to the establishment of a new port, and attempts at "beautifying" the city. These aspirations can be seen in the all-too-familiar clamping down on street traders and other ongoing "urban renewal" processes. They reinforce and create new precarious conditions that often painfully align everyday life inside and outside the popular neighbourhoods of the city with broader (and often global) urban trajectories, circulating policy mobilities and moments of heightened political repression (Branch and Mampilly 2015).

Urban policy orientations across Kampala have in recent years been directed through a city-regional structure in the form of the Kampala Capital City Authority (KCCA) (Goodfellow 2013; Gore and Muwanga 2014). The KCCA has increasingly focused on projects that "try to appeal to Uganda's urban middle class and elite, managing the city in their interest" (Branch and Mampilly 2015: 146). Yet 60 per cent of Kampala's residents live in over 30 informal settlements across the city, with 39 per cent living below the poverty line (Uganda Bureau of Statistics 2007). Life in the city, for many, is a shared experience of a lack of basic service provision, the ongoing threat of evictions, multi-faceted land and resource conflicts, and a sense that the dominant renderings of the city's future and indeed citizenship within Kampala are not for them.

Namuwongo is a centrally located popular neighbourhood to the south-east of the city centre. It is an industrial zone, partly built on a wetland and adjacent to an unused railway track leading to Port Bell (the port is one of the important elements of the redevelopment plans, although the work is yet to start). It has become a dense and diverse neighbourhood that sustains many different trajectories of urban life. In a city in which newcomers account for over 50 per cent of the total population (Lwasa 2011), Namuwongo has become a crucial space in which migrants are able to find cheaper sources of housing, connections to kin and economic opportunities – a popular neighbourhood.

Until the early 1980s, Namuwongo was dominated by mud and wattle structures (Mann and Andabati 2014). "Slum upgrading" work by the Government of Uganda and UN-Habitat began in 1984, leading to parts of the area above the railroad tracks being formalised and divided into plots. Many of the original beneficiaries sold their newly acquired land title or were marginalised from the process (Mann and Andabati 2014). The population increased due to migration from across Uganda and beyond, often driven by conflict and insecurity, and poorer residents moved to the wetland and railroad tracks. Today, an estimated 15,000 people live in Namuwongo (Mann and Andabati 2014), with few services and infrastructure (such as road access), leading to patchwork combinations of formal and informal infrastructures such as energy and waste (e.g. Brown 2015 on sanitation) that show the hybrid (Smith 2018) and heterogeneous conditions (Lawhon et al. 2018) through which infrastructure operates and is experienced.

The occasional enforcement of building controls within ten to fifteen metres of the railway track (see Figure 2.1) and wetland locations has meant that the threat of demolition and eviction continue to cast a long shadow over the future of the area (Kagenda 2014). In a household survey previously undertaken in Namuwongo (Silver 2015), nearly 80 per cent of participating residents are fearful of their dwellings being demolished and over 70 per cent of resident households consider their dwellings to be semi-permanent or temporary (Silver 2015). With partial demolition in 2014 by the KCCA and Rift Valley Railways near the tracks (aiming at a 30-metre clearance) and ongoing rumours concerning enforcement by the National Environment Management Agency (NEMA) of the wetlands, the only certainty is that the uncertainties and rumours are not going to go away. Given this backdrop, the few attempts at development in Namuwongo have tended to be small scale and led by NGOs such as Hope for Children that have built facilities to address issues like waste disposal, including open defecation (Brown 2015).

Figure 2.1 Enforcement of building controls near the railway

The aim of our research was to examine the ways in which different residents in Namuwongo navigate the multiple topographies of city life, where "topography" refers to the material geographies of the city in its different forms (routes, markets, public spaces, transport hubs, workplaces, home etc.). Between January and June 2015 we worked with local researcher and resident Joel Ongwec to recruit resident participants, and undertook situated fieldwork that created cartographies of urban navigations. It built on longer term engagement with the area and previous research (see Silver 2015) Recruitment was based on Joel's existing, extensive networks and was predicated on a need to find a sample that would reflect the diversity of Namuwongo in terms of age, gender, income, time in Namuwongo and where residents might have migrated from within Uganda. Over several research trips covering a period of seven months, working with six residents meant we were able to conduct repeated interviews, follow-along mini-ethnographies, focus groups, workshops with the residents and civil society groups, and work with the residents to develop and reflect on the process of exhibiting the six stories at an event in the Uganda National Museum. During all of this we talked with other residents too, conducted a larger survey and drew on a range of views as we progressed. The focus on six residents reflected some of the diversity of the neighbourhood: two widowed women with children to support (one younger (Josephine) and one older, with grandchildren (Jennifer)), one late teenage man who had lost his parents and depended on extended family (Amiri), two men in their late 30s who had young families but had different levels of income security (Ali and Isamail), and one older man who was economically the most secure and politically well-connected (Masengere). This is not a "representative sample" of this incredibly diverse neighbourhood, and caution needs to be used in building generalisations about life in Namuwongo from a small group. Nonetheless, the stories that emerge allow us to identify how different residents mobilise and use practices that enter into the making of everyday life in the area on the margins of citizenship.

Residents in Namuwongo often weave together a range of skills, knowledges and economic pursuits, and the six we worked with reflect some of that diversity. Jennifer has worked in many different ways to earn income, from bead-making to catering. She has also set up a women's group and helps local families apply for scholarships to schools. Just as there has been an uncertainty in her economic life, so too has there been insecurity at home and she has moved away from the house she proudly built herself. Ali, the second resident, moved to Namuwongo to be with his sister when he first arrived in Kampala from eastern Uganda. He started by selling fruit and eventually established one of the most popular stalls at the neighbourhood market. Ali's fruit and vegetable stall at the entrance to Namuwongo Market is an economic investment first and foremost, but it is also a lively social space where people chat, swap stories, laugh, debate and discuss the latest gossip. Before dawn, Ali travels to the city centre's Nakesero Market to pick up supplies from traders. He then quickly travels back to his stall in Namuwongo and will be there until after 9pm.

Josephine moved to escape insecurity and conflict in northern Uganda. She has benefitted from informal contacts with women in the neighbourhood that have helped her with selling fruit and vegetables in the city, a difficult and precarious livelihood with little income. Not having a reliable income is immensely challenging, especially when everyday needs – charcoal, paying for toilets, buying basic foods – take up almost all daily earnings. The Pentecostal Church has helped with school fees for her family. She and her neighbours often help each other, from looking after each other's children to finding a reliable bricklayer to repair the house after a storm.

Isamail sadly passed away in May 2015 due to poor health, at the age of 36, leaving behind his wife and four children. He was a matatu (privately owned mini-bus) driver who travelled the route between Namuwongo and the city centre. He started very young as a taxi conductor, and later became a driver. The matatus are important and cheap forms of travel for residents, but work was not always certain for Isamail. He had no guarantee that a matatu would be available for him to drive every day, or if he'd be well enough to work. If he did work, the driving involved long, tiring hours often stretching between 5am and 10pm. Like many residents, Isamail struggled to take care of his health, pay for medical treatment, provide for his family and maintain a steady income, reflecting the few benefits open to poorer citizens of Uganda.

Masengere was a chief in his village in western Uganda and is now a chairman in Namuwongo and an elected official for his part of the neighbourhood. He plays a vital political role in the area, and with his stamp can provide the approval needed for a wide range of local issues or economic pursuits from job references through to identity documents. In 1991 he was elected to the position of chairman and has since helped to create facilities in the neighbourhood. He believes that in time Namuwongo will become very different as part of a modernising and beautifying city, but hopes that he will be able to carry on grazing his few cattle on the nearby wetlands because it helps him feel connected to his rural home.

Finally, Amiri is proud of Namuwongo and its youth, having lived here for most of his life. He is young and stylish, and understands the challenges many young people in the area struggle with. Amiri lost both parents when he was young, and school didn't work out well – he finished early and started to earn money by collecting and selling scrap materials in and around the local area. With the help of his uncle, he was trained as a carpenter, producing an impressive range of different wood products in his uncle's workshop at the edge of the neighbourhood. Sometimes he leaves Namuwongo to work on sites across the city, including recently the large new Hilton hotel being built in the city centre. He lives with friends in a tiny, rented dwelling in the "Soweto" section, the densest and one of the poorest parts of the neighbourhood.

Collectively these residents experience citizenship on the margins, in constant negotiation with the state and the responsibilities it is supposed to fulfil in terms of the basic infrastructures of social reproduction in cities from housing through to networked services.

Social infrastructure

"Infrastructure" has become a dominant lens through which to conceptualise the urban condition and research urban life. "Social infrastructure", however, has received less attention especially in relation to citizenship (see Lemanski in this collection). It is not just a context or a noun, but a verb: social infrastructure is made and held stable through work and changing ways of connecting. It is a connective tissue, often unpredictable, anchoring urban life in popular neighbourhoods across the urban world (e.g. see De Boeck 2015; Graham and McFarlane 2014; Silver 2014; Simone 2014), varying in form and content.

Our use of the term "social" in "social infrastructure" is designed to underline that these infrastructures are "peopled", to echo AbdouMaliq Simone's (2004) influential arguments in his wonderful piece, "People as infrastructure". The terms "network" or "social capital" are less useful here. We are not concerned with narrow valuations of social interactions, nor are we solely focused on the economic potentials of social collectives (Putnam 2001). Our focus is on the wider sets of ways in which people manage and navigate the city on the margins of citizenship. Moreover, infrastructure is more effective at capturing flows, movements, congestions and internments of people and things. If Putnam's (2001) notion of "social capital" emphasises the collective potential to bring economic and political change, then the term social infrastructure is more focused on social reproduction across the everyday conditions of inequality in cities such as Kampala. It is a survival strategy, rather than a process of empowerment.

In this reading, social infrastructure becomes a way to stretch long-standing work on urban poverty and "assets" in popular neighbourhoods to account for the movement and geographies of life in and beyond the margins of citizenship that span the urbanisation process. For Caroline Moser (2010), local assets are vital resources through which residents in popular neighbourhoods are able to cope with and perhaps even move out of poverty. These include physical (e.g. infrastructure, resources), financial (e.g. savings, credit), human (e.g. education, health), social (e.g. norms, reciprocity) and natural (e.g. land, environmental conditions) capital, and it is through household and community accumulations within and between these assets that people move from vulnerable to more stable positions (see also Banks 2016; Bebbington 1999).

Care is a crucial social infrastructure in Namuwongo in thinking about the ability to survive the city. Moments of caring are visible throughout the participants' everyday lives as friends, family and neighbours come together to secure essential needs, sharing and supporting each other during heightened moments of crisis and the challenges of each day on the margins. Each of the six residents both called on and enacted these social infrastructures in different ways, reflecting their capacities, backgrounds, networks, experiences and empathies.

Through past experience as an elected chairwoman in the nearby town of Jinja, and her struggles as a widow to sustain a family and links to a range of individuals and organisations inside and outside the neighbourhood, Jennifer has found herself at the centre of numerous moments of care that make a substantive difference to many people. Jennifer has worked with the local Pentecostal Church to build sanitation facilities for collective use by the congregation, helps out her family members to start businesses, and seeks out sponsorship for Namuwongo children to attend school. And then there is the more informal work of checking up with people, seeing how they are getting on, identifying problems and providing contacts to help – much of this picked up while moving through streets and lanes of the neighbourhood.

These relations have been in large part developed through her prominent role in a women's group, which includes the making and selling of crafts, and a savings group covering everything from everyday essentials such as food or fuel when times are tough through to exceptional needs such as funeral costs and, in some cases, saving towards a small business. These social infrastructures of care are at once social and material, ranging from food and fuel to craft skill training, basic services, school access and contacts for informal employment.

Care should not necessarily be seen as altruism. Jennifer has positioned herself as locally important, and as such often benefits in a range of ways, from small donations from parents for whom she has secured a school place, through to selling her handmade beads to local and international contacts. The time and work that Jennifer puts in to creating and sustaining social infrastructures of care are also an investment in herself and in a wider reciprocity that is at once ethical and economic. This story of reciprocity repeats-with-difference in popular neighbourhoods globally. For example, reciprocal exchanges form the basis of insurance systems through which people borrow, lend, buy or sell between themselves, as has been documented in classic anthropological works such as Carol Stack's (1974) *All Our Kin*, which highlights the dense networks of reciprocity, debt and circulation of objects, favours and services in poor neighbourhoods (also see Gill 2000; Moser 1996).

Social infrastructures of care vary both between and within cities and dependent of the citizenship status of various residents (Brown et al. 2010). For example, Josephine's caring infrastructures are far less extensive than Jennifer's, yet they are all the more important because Josephine and her children are less able to secure the basic necessities of survival. Social infrastructures enable Josephine to work, for instance as neighbours look after her children, and at times of intensified crisis social infrastructures are absolutely vital for her survival. In situations of crisis the social infrastructure can spatially widen, including to rural family hinterlands, pulling in distant family alongside neighbours and friends. For example, on one occasion while selling fruit in town she seriously injured her leg, and family from her village in the north of Uganda sent food, while locally neighbours and friends helped when they could.

Again, there is a measure of reciprocity here: Josephine too will occasionally look after the children of neighbours, and she had also taken responsibility for the care of her niece, who has HIV and was unable to secure treatment in the village where she had lived, showing that the city may offer more than the rural. This meant another mouth to feed in a context where food is often scarce and where Josephine can seldom scrape together the money to get her children into school, placing yet more pressure on her and making the social infrastructures of care all the more vital.

Or take the example of Amiri. His relative security in his carpentry job is assured through his uncle. He has moved in and out of essential social infrastructures of care, from his early experience of losing his parents and becoming a waste-picker and later a carpenter. Amiri spoke of a growing realisation that he lives in an urban space in which caring infrastructures are necessary, whether in finding employment through contacts, or to look out for each other when the police are in the neighbourhood. If conditions for Amiri stay the same, and if he is able to continue helping his friends, then he too may well become a node in the infrastructure of care, as well as being someone who is beginning to move *beyond* a dependence on infrastructures of care (see below). While Jennifer is able to draw on a lifetime of experience to construct such infrastructures, Amiri is aware too of the potential and limits of constructing and sustaining social infrastructure.

These sets of socio-technical relations show the multiple ways through which social infrastructures are produced by marginalised residents in Namuwongo as a survival strategy, as a method of navigating the city in the present, into the future and as an everyday response to the failure of the state to provide the necessary service provisions required for social reproduction, and they are understood as a crucial comment of citizenship.

The limits of social infrastructure

We started to think of everyday life in Namuwongo dialectically precisely because we found ourselves returning to the limits of social infrastructure. There is no starker limit on social infrastructure or example of a lack of full citizen rights than eviction and demolition. Jennifer witnessed the late night demolition and eviction of homes by the authorities in 2014. Soon after, she relocated to a safer space in the neighbourhood. Her relocation outside of the designated zone of demolition protects the infrastructures she has worked so hard to develop over the years, despite the loss of close association with her neighbours, the effort in building the house, the costs of moving and subsequent renting and the peripheral location of her new home in the neighbourhood. The experience of displacement in popular neighbourhoods unfolds in distinct ways for different residents (also see Gillespie 2016). If citizenship is about the relations between state and citizen in which both have rights but also responsibilities (Cornwall and Coelho 2007) then the demolitions show that the Ugandan state is failing this relationship through its failure to provide housing and serve provision, and the punitive capacity to demolish the housing of people like Jennifer.

Beyond demolition, everyday interactions with the state expose residents to harassment and intimidation, often focused on people's efforts at generating income. Josephine faces the harassment of KCCA officials tasked to stop any street trading in the central city. While some days might involve helpful support and advice from other sellers about the movements of municipal enforcement teams, other days may leave her vulnerable, and occasionally she will drop what she's selling and rapidly move for cover when she sees the dreaded yellow shirts in the distance. This is the nature of everyday urbanism for many of Kampala's street traders as they navigate what Lindell and Ampaire (2016) term, after Oren Yiftachel (2009), "grey space". As Brown et al. (2010: 666) argue, "the increase in street traders has not been accompanied by a corresponding improvement in their status as citizens." Meanwhile, the ongoing harassment of youth in Namuwongo by the police, common across many popular neighbourhoods (Kimar 2014), limits Amiri's social infrastructures. While he is able to call on his friends, there is little that they can do to stop the police detaining them and demanding money for release. For Josephine and Amiri, and in relative contrast to Jennifer or Masengere (see below), the dialectical production of everyday urbanism often leaves social infrastructure as a weak resource.

Another limit to social infrastructure is crime. A house robbery soon after his arrival in Namuwongo meant that Ali lost much of the capital that he had built up during the early days of his business and showed how the state failed to fulfil its policing responsibilities toward its citizens. While social infrastructure kicked in here, in that his friends and family rallied around to offer emotional and financial support, the realities of crime and absence of security in a popular neighbourhood such as Namuwongo remain an ongoing threat, despite the introduction of policing units to the area. Many residents, despite knowing neighbours intimately, are not prepared to walk to nearby toilet blocks at certain times, especially at night, illustrating how various citizen rights from safety to accessing sanitation are not guaranteed by the state. Social infrastructures cannot provide adequate protection from crime and high levels of violence against women (Viswanath and Mehrotra 2007). Navigating the city is not just about knowing and mobilising various forms of social infrastructure to get by and get on, but also about being aware of the limits of these forms of support, to know when and where one is moving. We should, of course, be very wary of representations of "slums" as inherently dangerous spaces in relation to the wider city (Lombard 2015), but it would be wrong to deny that the threat and actuality of violence of different sorts is often more common in popular neighbourhoods given the desperate poverty that sometimes exists (Moser and McIlwaine 2014).

Finally, social infrastructures of care are both undermined by and emerge as vital through the violence inherent in different forms of policy. Arguably most destructive have been the structural adjustment programmes (SAPs) imposed by the World Bank/IMF, partly responsible for years of under-investment in Kampala (Branch and Mampilly 2015) and a key factor in the failure of the state to provide the rights of citizenship. This history inevitably meant that a disparate tissue of civil society groups, missionary enterprises and private practitioners – alongside and often interacting with the social infrastructures we describe here – have filled the voids. This is most tragically visible in the ways in which the Uganda health sector has been under-invested in for decades (Pfeiffer and Chapman 2010) and particularly in the challenges faced in relation to the HIV/AIDS epidemic (see Wallman 1996). SAPs pushed fee-driven government health systems, resulting in the devastating impact of HIV/AIDS due to the poorest being unable to access treatment (HIV prevalence increased from 6.4 per cent in 2004/5 to 7.4 per cent in 2012/13, and is higher in urban areas and among working-class groups; Uganda Country Progress Report 2014). Isamail was unable to secure the life-saving drugs and treatment needed after a series of complications and periods of bad health. Here the limits to both citizenship and social infrastructure are made visible in the most profound ways possible and the precarious nature of life in Namuwongo exposed at its most bare.

Set against these challenges, social infrastructures are vital forms of life support. But they are supplemented by other practices that are important to the making of everyday life and ways of navigating the city. *Consolidation* is the differential capacity to enhance position, while *coordination* is a role that emerges from consolidating position. These practices move beyond social infrastructure, but they are related in that they can provide enhanced capacities that may help shape social infrastructure of care. *Speculation* is an orientation to urban life itself, which is not determined by social infrastructure, consolidation or coordination, but which can be enhanced by them. Social infrastructure, consolidation, coordination and speculation are forged dialectically in relation to one another and the wider city: they co-evolve, sometimes reinforce, sometimes contradict, and emerge in distinct ways for the different residents.

Urban capacity: consolidation, coordination, speculation

Social infrastructure is a differentiated set of socio-material practices that make provisions for some of the basics of urban life: being able to get to work, to access schooling, to get basic healthcare, and so on. Consolidation, in contrast, refers to a greater level of economic security, a position where social infrastructures do not cease to matter, but where the conditions of everyday life have exceeded those infrastructures, where people are less dependent on them, even as consolidation can provide the physical spaces through which to enhance their capacities to develop those social infrastructures.

Of all the residents, Masengere was the one who was the most consolidated. His position in the wider area is relatively secure, ensured by his role as a local party Chairman for the ruling branch of the country's dominant political force, the National Resistance Movement (NRM), and the financial benefits, political and social contacts, and reputation the role brings. Indeed, such is his relative security that he was the least concerned about possible future development of all the respondents, a security added to given that he can use political power to manoeuvre into a better position than most and the fact that he resides outside demolition target areas. Still, even his consolidated position has its physical limits; he believes, for instance, that the area will eventually become too expensive for current residents and that KCCA will have to pay compensation to those who will lose their homes.

Amiri learned skills that helped bring a measure of security through training from his uncle in woodwork and this is physically consolidated by the ability to access the business premises and the customer networks that have been established around it. The more experience and skills Amiri gets of different kinds of woodwork through the workspace of his uncle, the more he secures his position in the present and into the future. But it would be wrong to give the impression that Amiri was simply in a position of growing security and confidence, or that learning skills is a straightforward ticket out of poverty. Uncertain futures press up against consolidating practices and the spaces established to support these strategies. Amiri is aware of how the city's inequalities limit opportunities, and he alluded to a vaguely felt threshold in the capacity to consolidate further: "I sometimes feel positive about changing things, but then I think no: I cannot change anything". To some extent, this is about the KCCA and the limits it puts on urban opportunities – "if you have a small plot of land the KCCA will say it doesn't meet standards and move you on" – and about the impact of urban change more broadly: "All of the changes made have hurt someone. Change is never smooth". He is acutely aware of a lack of jobs for young people and the exclusions they face in accessing urban spaces of opportunity and accumulation. The dialectics of everyday urbanism are forged as much through what is *absent* as through what is present.

Amiri's position reflects the wider "intergenerational bargain" through which young people's life chances are supported by reciprocal responsibilities of care from adults (Day and Evans 2015). As stable employment opportunities recede for many young people in increasingly competitive and precarious global labour markets, family care can be vital or even the only forms of support for a "lost generation" of youth that Honwana (2012) has characterised as remaining in a state of "waithood" (see Jeffrey 2010). Uganda has the highest poverty incidence among young people (between 18 and 30) in sub-Saharan Africa, with 94 per cent living below US$2 per day in a society in which 78 per cent of its population are under 30 years of age. This situation in Uganda reflects the broader urban African experience for many young people, left shunting repeatedly from limited social infrastructures of care to efforts to consolidate some measure of often precarious economic security.

For Ali, the ability to work with customers, build rapport or know how to negotiate to buy the best products for the right prices to sell in his stall are all capacities that have helped him consolidate a relatively stable economic position. His vision for Namuwongo revolved around growing business opportunities, including connections to the rest of the city and its hinterlands. He was, for example, enthusiastic about the new road KCCA had built through Namuwongo. Indeed, he connected the construction of the road, in the same breath, to a more secure neighbourhood – a big concern for Ali, given that he was once robbed: "KCCA has constructed the road in Namuwongo to Bugolobi, and the problem of sanitation and robberies has stopped". It follows that the concerns he had about this consolidating position was the stability of the market. In a neighbourhood with frequent demolitions and uncertainties about who owns what land, it is hardly surprising that there are also rumours about the market being demolished. At one workshop, Ali was partly reassured by an NGO representative that this was only rumour and that the market's future was supposedly secure.

Ali was in quite a different position from Isamail and Josephine in this respect. Isamail struggled to consolidate his position due to poor health and the unreliability of available taxis when he was well enough to drive, and so his income was far more unpredictable and his ability to ensure provisions of food, water, schools fees and other costs for himself and his family were more curtailed. Kampala, for Isamail, was less a space of consolidating and more an exhausting struggle of negotiating traffic, knowing how and when to pay the municipal officials a little to jump the queue, working out where the short-cut mud roads are and when to take them, and being at the taxi park (also see Goodfellow 2015). From the age of 13, his life had revolved around the main taxi park, with friends, waiting and talking, but consolidation largely eluded him. He was planning to start a retail shop selling food in an attempt to find other ways beyond depending on limited social infrastructures. In the end, not only was Isamail far from consolidating any security, even social infrastructures of care were not enough. His death has made it harder still for his young family to consolidate into the future and access the rights of citizenship. As Day and Evans (2015) show in their work on familial care in Zambia, the loss of a parent can accentuate other difficulties, for instance with children often then dropping out of school, having to spend more time on household labour and caring duties for siblings – especially for girls – or migrating to live with relatives.

For Josephine, for whom everyday life is about survival and being able to spend adequate time in the city to sell enough fruit to put food on the table for her children, niece and herself, the potential to consolidate exists largely as a dream. Josephine's income is unpredictable and almost entirely spent on basics: charcoal for cooking, toilets, food, maintaining the house and school fees if possible. The KCCA placed a severe restriction on Josephine's hopes of consolidating, and she connected the municipality to gender, opportunity and urban space: the "KCCA does not find a way of collaborating with women who sell goods in town. They don't think of the women". Josephine's experience of the punitive state echoes the experience of demolition faced by Jennifer and in which the responsibilities of the Uganda government in relation to citizenship are pushed aside in favour of advancing its capacity to punish citizens and police urban space that create gendered inequalities. As Gina Porter (2011) has shown, if poor provision and maintenance of roads, transport services and safety are concerns across many African contexts, these issues are exacerbated for women, and further still given the cultural constraints on women's mobilities. Josephine is developing plans, however, and spoke of setting up a business to buy fish in the town and sell it back in the village.

These practices of consolidation are highly differentiated. They are not consolidation to the extent that Moser (2010) identifies in her work on low-income neighbourhoods in Ecuador, in which the built environment is consolidated alongside improved livelihoods and education. And while for some residents consolidation is rare or perhaps a never-achieved hope, there are those whose capacities have gone beyond collaborating with others to a more organisational role of *coordinating* different economic, social and political processes.

Coordination and speculation

Coordination is the capacity to organise social, economic or political activities in the neighbourhood, and emerges through a more consolidated position in the neighbourhood. Coordination can take the form of enacting a social infrastructure of care, or can influence other trajectories in the neighbourhood, and like care and consolidation it too is radically differentiated across the residents.

Jennifer, for instance, sometimes coordinates people and jobs, or brings women recently widowed into the informal social support of other women, or links up parents and schools. Masengere, meanwhile, coordinates activities characterised more by profit and prestige through his role as a key community gatekeeper to officialdom, through which he has the power to allow or block developments in the area. Compare this to Josephine, whose struggle just to survive leaves her with little capacity to coordinate urban life. Still, there were some coordinating roles for Josephine, for example in sharing information with other women in the city selling fruit about the location of KCCA officials, or about where they did or did not have much success selling fruit. Isamail was in a similar position to Josephine, in that his health prevented him getting into a position of consolidation that would then open out the possibility of coordination. He was, however, helping to pay for his younger brother to be trained as a lawyer, and in this way coordinating family members and opportunities in the limited way that he was able to. Ali, in contrast, coordinates food through the market, neighbourhood, city and beyond.

Alongside consolidation and coordination is *speculation*. Speculation is a practice of imagining and acting into the future and it occurs through various moments of calculation about how to navigate the city. It can refer to everyday and relatively limited forms that may influence the success of a day, through to longer term speculating around livelihoods and material arrangements that involves potential risk (Simone 2014; Zeiderman et al. 2015) but also shows the type of imageries through which residents' hopes and aspirations are materialised. It is a way of opening out opportunities and capitalising on changing circumstances, and depends on the socio-economic position and perceptions that different residents bring to the often turbulent currents of everyday life. Namuwongo is made up of many migrants, whose trajectories are orientated around building contacts, skills and spaces needed to navigate urban life. The neighbourhood presents a set of densities that feature not just as challenges, but opportunities for the immediate present and future (Wallman 1996).

There is always a risk with speculation, a chance that an investment of money or time or hope might not pay off, that the expected outcomes may fail to materialise and that the resident is left worse off. Ali, able to quickly acquire knowledge about the food business and the market in Namuwongo from the other traders who discuss, exchange and reflect on how the place operates according to perceived principles, rules and rhythms, makes constant, long- and short-term speculations in the hope of enhancing his livelihood. For example, he decided to pay an extra 10,000 shillings a month (£3) to acquire the front stall, visible on entry to the market from its main entrance, in order to interact with a wider number of customers and therefore increase trade. The decision paid off and his turnover increased. Josephine performs a different kind of speculation in that each day she makes small but important decisions about which part of the central city to work in and which parts to avoid and when (in relation to KCCA officials), bearing in mind the distributions of customers and urban flows. For Josephine, speculation operates on registers of anxiety and hope, as opposed to the relatively more secure business decisions made by Ali.

We have shown that social infrastructure is an essential coping mechanism that has its limits for the residents of Namuwongo, with practices of consolidation, coordination and speculation mobilised to address the challenges of living on the margins in Kampala. We have emphasised the importance of understanding citizenship through these everyday practices as essential for the social reproduction of various households and in myriad ways. The capacities of different residents to develop these practices shows the ways in which different households possess different capacities to deal with and develop responses to the topographies of urban life.

Conclusion

Social infrastructure, consolidation, coordination and speculation are vital and closely inter-related processes through which residents navigate Namuwongo and the wider topographies of Kampala, and the ways in which they cope with and seek to move beyond the limited forms of citizenship they experience. Without these processes it is difficult to see how life in this often precarious, uncertain context could be reproduced. They operate not separately but dialectically, shaping one another, placing limits and forging different assemblages of urban life for different residents that open up important considerations concerning the geographies of social infrastructure and the shifting terrains of everyday urbanism.

Jennifer's centrality to social infrastructure enhances her ability to consolidate, while Josephine's limited access to social infrastructure is both vital but also reflective of the struggle she has to consolidate, coordinate or speculate in the ways that, for instance, Ali can. The limits placed on Isamail's social infrastructure by the political economies of health, services and infrastructure left him bereft of life-sustaining support. His sad death is not an uncommon story in the neighbourhood, and the consequence of course is that his wife and children are all the more dependent on social infrastructures, all the more limited in their chances to move beyond them into consolidation or coordination. Amiri, in contrast, was rescued by his social infrastructure of care in the form of his uncle, and he inhabits a quite different assemblage that opens out a potential future in which he increasingly consolidates and perhaps coordinates, for example in relation to youth groups or training other youngsters in woodwork. Masengere was the most secure of all respondents; with relatively little need for social infrastructures of care, he is well consolidated and is a key coordinating node in the area, including in other people's social infrastructures and capacity to consolidate or speculate.

The dialectical relationship between the city, state, citizenship and everyday life in popular neighbourhoods is crucial to the ways in which urbanism is produced and how poverty is shaped and navigated. In Namuwongo, the visible gaps in state provision of health, education (spending on which more than halved between 1998 and 2002) and physical infrastructure show the need to research the ways in which residents navigate the city beyond the security of rights or citizenship. It means residents either expend enormous energies to make up for these deficits and/or are left unable to secure basic provisions. It is at these moments, in which basic rights of citizenship are barely visible, that forms of collaboration with the state might become vital in providing the types of support and opportunity that are so difficult, and in many cases impossible, through the efforts of residents individually and collectively. This is complicated by the dual role of the state: on the one hand, the KCCA, alongside the wetland and railway authorities, are planning to evict and demolish many parts of Namuwongo, partly in view of a major redevelopment plan for the city that extends to Lake Victoria (Guma 2016). But on the other hand, the municipality has also been responsible for providing at least some essential urban services – even if they are partial and temporary – such as waste collection and road improvements. The state too is dialectical, and the consequence is an uncertain future for the neighbourhood. It is these conditions through which urban scholars should develop broader conceptions of infrastructure. The chapter has shown that conceptions of urban infrastructure must be pushed beyond the hardware, technologies and networks that have traditionally been its focus and toward a broader conception that emphasises the dynamic relationship between citizenship, the failure of the state to provide service provision and the everyday actions through which communities such as Namuwongo are able to socially reproduce. Here we wish to highlight the ways in which physical infrastructure (or the lack of it in urban spaces such as Namuwongo) becomes an important lens to open up a series of ways of approaching how we understand and examine everyday urban life.

A fundamental challenge lies in the state's unwillingness to engage with their citizens, their experiences and perceptions, and their potential to help remake the city. This illustrates both ongoing national political tensions (Branch and Mampilly 2015) and also the ways in which communities like Namuwongo are stigmatised and marginalised by the state and private-sector partners (Wacquant 2008). One of the most striking features of the research process was that the residents we spoke with were eager to contribute as citizens to the development of their neighbourhood and to have a voice in urban planning in Kampala more generally, as other (more secure) popular neighbourhoods are doing. Amiri spoke about forming or becoming involved in youth groups, and Jennifer and Josephine spoke about the need for training groups for widows and organisations to support children into schools. Ali spoke about setting up an association that could represent the interests of the market to the KCCA, and which would perhaps generate some security in an area where the rumour of demolition is never far from the surface. There was talk about setting up a forum of some kind that would bring together these different issues, ideas for action and opportunities for interaction between residents, the state and NGOs. There were questions raised about how the city might be redeveloped with the aspirations of ordinary residents, rather than with the interests of the middle and upper classes (Branch and Mampilly 2015), more centrally in mind. Ambitious plans, to be sure, for residents so often squeezed in their time and resources.

The future for Namuwongo is profoundly uncertain and precarious, as it is for so many popular neighbourhoods globally in which citizenship means little for people living on the margins (Brown et al. 2010: Holston 1999: Rao 2010). At least part of the role for critical urban scholarship is to understand how everyday life is shaped on the margins of citizenship through social infrastructure and its limits. Doing so reveals, for example, that both the challenges of urban life and the solutions to them are radically differentiated even within a neighbourhood. Yes, there are some staples of citizenship that should be available to all: decent secure housing, water, sanitation, energy, education, healthcare and opportunity, all of which have featured in our account. But the ways in which these are developed and delivered often requires an understanding of the diverse needs and trajectories of the actually existing urban geographies within and beyond the neighbourhood. Policy, practice, activism and scholarship are all more effective with a stronger grasp of how they might operate through social and spatial difference, and with people in the lives and neighbourhoods they invest in.

Acknowledgements

This chapter is a based on a revised paper McFarlane, C., and Silver, J. (2017). Navigating the city: dialectics of everyday urbanism. *Transactions of the Institute of British Geographers.* (http://onlinelibrary.wiley.com/doi/10.1111/tran.12175/full). The authors would like to thank the publisher and journal editor for allowing re-use. The research was funded by the Leverhulme Trust. Research support was provided by Joel Ongwec.

References

Banks, N., 2016, Livelihoods limitations: The political economy of urban poverty in Dhaka, Bangladesh. *Development and Change* 47(2): 266–292.

Bayat, A., 2010, *Life as Politics: How ordinary people change the Middle East*, Stanford University Press, Redwood City, CA.

Bebbington, A., 1999, Capitals and capabilities: A framework for analyzing peasant viability, rural livelihoods and poverty. *World Development* 27(12): 2021–2044.

Benjamin, W., 2003, *The Arcades Project* (H. Eiland and K. McLaughlin trans), Harvard University Press, Cambridge, MA.

Branch, A. and Mampilly, Z., 2015, *Africa Uprising: Popular protest and political change.* Zed Books, London.

Brown, A., Lyons, M. and Dankoco, I., 2010, Street traders and the emerging spaces for urban voice and citizenship in African cities. *Urban Studies* 47(3): 666–683.

Brown, S., 2015, Kampala's sanitary regime: Whose toilet is it anyway? In S. Graham and C. McFarlane (Eds.), *Infrastructural Lives: Urban infrastructure in context*, Routledge-Earthscan, London: 153–173.

Caldeira, T.P., 2000, *City of Walls: Crime, segregation, and citizenship in São Paulo*, University of California Press, Oakland, CA.

Comaroff, J., & Comaroff, J.L., 2015, *Theory from the South: Or, how Euro-America is evolving toward Africa*, Routledge, London.

Cornwall, A. and Coelho, V.S. (Eds.), 2007, *Spaces for change? The politics of citizen participation in new democratic arenas*, Zed Books, London.

Day, C. and Evans, R., 2015, Caring responsibilities, change and transitions in young people's family lives in Zambia. *Journal of Comparative Family Studies* XLVI(1): 137–152.

De Boeck, F., 2015, Divining the city: Rhythm, amalgamation and knotting as forms of urbanity. *Social Dynamics* 41(1): 47–58.

De Certeau, M., 2011[1984], *The Practice of Everyday Life*, University of California Press, Los Angeles, CA.

Engels, F., 1987, *The Condition of the Working Class in England*, Penguin, London.

Gill, L., 2000, *Teetering on the rim: Global restructuring, daily life, and the armed retreat of the Bolivian state*, Columbia University Press, New York.

Gillespie, T., 2016, Accumulation by urban dispossession: Struggles over urban space in Accra, Ghana. *Transactions of the Institute of British Geographers* 41(1): 66–77.

Goodfellow, T., 2013, The institutionalisation of "noise" and "silence" in urban politics: Riots and compliance in Uganda and Rwanda. *Oxford Development Studies* 41(4): 436–454.

Goodfellow, T., 2015, Taming the "rogue" sector: Studying state effectiveness in Africa through informal transport politics. *Comparative Politics* 47(2): 127–147.

Gore, C.D. and Muwanga, N.K., 2014, Decentralization is dead, long live decentralization! Capital city reform and political rights in Kampala, Uganda. *International Journal of Urban and Regional Research* 38(6): 2201–2216.

Graham, S. and McFarlane, C. (Eds.), 2014, *Infrastructural Lives: Urban infrastructure in context*, Routledge, London.

Guma, P.K., 2016, The governance and politics of urban space in the postcolonial city: Kampala, Nairobi and Dar es Salaam. *Africa Review* 8(1): 31–43.

Hall, S., 2012, *City, Street and Citizen: The measure of the ordinary*, Routledge, London.

Hall, S.M., 2015, Migrant urbanisms: Ordinary cities and everyday resistance. *Sociology* 49(5): 853–869.

Hansen, T.B. and Verkaaik, O., 2009, Introduction–urban charisma: On everyday mythologies in the city. *Critique of Anthropology* 29(1): 5–26.

Holston, J., 1999, *Cities and Citizenship*, Duke University Press, Durham, NC.

Holston, J., 2008, *Insurgent Citizenship: Disjunctions of democracy and modernity in Brazil*, Princeton University Press, Princeton, NJ.

Honwana, A.M., 2012, *The Time of Youth: Work, social change, and politics in Africa*, Sterling,VA, Kumarian Press.

Jacobs, J., 1961, *The Death and Life of Great American Cities*, Vintage, London.

Jeffrey, C., 2010, *Timepass: Youth, class, and the politics of waiting in India*, Palo Alto, CA: Stanford University Press.

Kagenda, P., 2014, KCCA's night raiders, *The Independent*, 18 August.www.independent.co.ug/news/news-analysis/9244-kccas-night-raiders (accessed 10 February 2016).

Kihato, C.W., 2011, The city from its margins: Rethinking urban governance through the everyday lives of migrant women in Johannesburg. *Social Dynamics* 37(3): 349–362.

Kimar, W., 2014, On raids and connecting favela resistances in Kenya and Brazil. http://blogs.lse.ac.uk/favelasatlse/ (accessed 2 August 2016).

Lancione, M., 2014, Entanglements of faith: Discourses, practices of care and homeless people in an Italian City of Saints. *Urban Studies* 51(14): 3062–3078.

Lawhon, M., Nilsson, D., Silver, J., Ernstson, H. and Lwasa, S., 2018, Thinking through heterogeneous infrastructure configurations. *Urban Studies* 55(4): 720–732.

Lefebvre, H., 1991[1958], *The Critique of Everyday Life (Vol. 1)*, (J. Moore trans), Verso, London.

Lindell, I. and Ampaire, C., 2016, The untamed politics of urban informality: "Gray space" and struggles for recognition in an African city. *Theoretical Inquiries in Law* 17(1): 257–282.

Lombard, M., 2015, Discursive constructions of low-income neighbourhoods. *Geography Compass* 9: 648–659

Lomnitz, L., 1977, *Networks and Marginality: Life in a Mexican shantytown*, Academic Press, New York.

Lwasa, S., 2011, Sustainable Urban Development: Managing city development in Uganda. In E. Birch and S. Wachter (Eds.), *Global Urbanization*, University of Pennsylvania Press, Philadelphia, PA: 276–293.

Mann, D. and Andabati, D., 2014, The informal settlement of Namuwongo: A baseline survey for the PECTIS Project. http://suda-africa.org/wp-content/uploads/2014/05/PECTIS_Baseline-Survey-Report_Validation-Draft_reduced.pdf (accessed 14 January 2015).

McFarlane, C., 2011, *Learning the City: Knowledge and translocal assemblage (Vol. 101)*, John Wiley & Sons, Hoboken, NJ.

McFarlane, C., and Silver, J., 2017, Navigating the city: Dialectics of everyday urbanism. *Transactions of the Institute of British Geographers* 42(3): 458–471.

Moser, C., 1996, *Confronting Crisis: A comparative study of household responses to poverty and vulnerability in four poor urban communities*, The World Bank, Washington, DC.

Moser, C., 2010, *Ordinary Families, Extraordinary Lives: Assets and poverty reduction in Guayaquil, 1978–2004*, Brookings Institution Press, Washington, DC.

Moser, C., and McIlwaine, C., 2014, Editorial: New frontiers in twenty-first century urban conflict and violence. *Environment and Urbanization* 26: 331–344.

Perlman, J.E., 1979, *The Myth of Marginality: Urban poverty and politics in Rio de Janeiro*, University of California Press, Los Angeles, CA.

Perlman, J.E., 2006, The metamorphosis of marginality: Four generations in the favelas of Rio de Janeiro. *The Annals of the American Academy of Political and Social Science* 606(1): 154–177.

Perlman, J.E., 2010, *Favela: Four decades of living on the edge in Rio de Janeiro*, Oxford University Press, Oxford.

Pfeiffer, J. and Chapman, R., 2010, Anthropological perspectives on structural adjustment and public health. *Annual Review of Anthropology* 39: 149–165.

Porter, G., 2011, "I think a woman who travels a lot is befriending other men and that's why she travels": Mobility constraints and their implications for rural women and girls in sub-Saharan Africa. *Gender, Place and Culture* 18(1): 65–81.

Putnam, R.D., 2001, *Bowling Alone: The collapse and revival of American community*, Simon and Schuster, New York.

Rao, U., 2010, Making the global city: Urban citizenship at the margins of Delhi. *Ethnos* 75(4): 402–424.

Roy, A., and Crane, E. (Eds), 2015, *Territories of Poverty*, University of Georgia Press, Athens, GA.

Roy, A. and Ong, A. (Eds.), 2011, *Worlding Cities: Asian experiments and the art of being global*, John Wiley & Sons, Oxford.

Silver, J., 2014, Incremental infrastructures: Material improvisation and social collaboration across post-colonial Accra. *Urban Geography* 35(6): 788–804.

Silver, J., 2015, *Survey of 150 Households in Namuwongo*, Working Paper (Unpublished).

Simone, A., 2004, People as infrastructure: Intersecting fragments in Johannesburg. *Public Culture* 16(3): 407–429.

Simone, A., 2010, *City Life from Jakarta to Dakar: Movements at the crossroads*, Routledge, New York.

Simone, A., 2014, *Jakarta, Drawing the City Near*, University of Minnesota Press, Minnesota, MN.

Smith, S., 2018, Hybrid networks, everyday life and social control: Electricity access in urban Kenya. *Urban Studies*, doi: doi:0042098018760148.

Stack, C., 1974, *All our Kin*, Basic Books, New York.

Uganda Bureau of Statistics, 2007, Nature, distribution and evolution of poverty and inequality in Uganda. www.ubos.org/onlinefiles/uploads/ubos/pdf%20documents/ILRI%20Poverty%20Report%202007.pdf (accessed 12 September 2015).

Uganda Country Progress Report, 2014, *The HIV and AIDS Uganda Country Progress Report, 2014*, Ministry of Health, Government of Uganda, Kampala, Uganda.

UN-Habitat, 2011, *Urban Profile of Kampala*, UN-Habitat, Nairobi, Kenya.

Vasudevan, A., 2015, The makeshift city: Towards a global geography of squatting. *Progress in Human Geography* 39(3): 338–359.

Viswanath, K. and Mehrotra, S.T., 2007, "Shall we go out?" Women's safety in public spaces in Delhi. *Economic and Political Weekly* 42(17): 1542–1548.

Wacquant, L., 2008, *Urban Outcasts: A comparative sociology of advanced marginality*, Polity, Cambridge.

Wallman, S., 1996, *Kampala Women Getting By: Wellbeing in the time of AIDS*, Ohio University Press, Athens, OH.

Watson, V., 2014, African urban fantasies: Dreams or nightmares? *Environment and Urbanization* 26(1): 215–231.

Yiftachel, O., 2009, Critical theory and "gray space": Mobilization of the colonized. *City* 13(2–3): 246–263.

Zeiderman, A., Kaker, S.A., Silver, J. and Wood, A., 2015, Uncertainty and urban life. *Public Culture* 27(2(76)): 281–304.

3 The politics of urban sanitation

Making claims to the city

Colin McFarlane

Introduction

The 21st century, we are continually told, will be urban. By 2030, 60% of the global population will live in cities. Currently, more than one in four urbanites – a billion people – reside in informal settlements. Informal settlements are urbanising faster than cities more generally, and by 2050 it is expected that three billion people – a third of the global urban population – will be housed in different kinds of informal urban spaces (Mitlin and Satterthwaite 2013). As both the 2015 UN *Sustainable Development Goals* (SDGs) and the 2016 *New Urban Agenda* make clear, informal neighbourhoods present an array of fundamental challenges for addressing poverty and environmental change, and are arguably the key challenge for cities.

Within informal settlements, a central arena for improving life conditions is sanitation. As informal settlements urbanise, so too does the global sanitation crisis. Almost 25% of the 2.6 billion people lacking adequate sanitation live in urban environments, mostly in informal settlements, and that proportion is growing. One fifth of the world's population is forced to regularly defecate in the open (Mara 2012). Diarrhoea, usually the result of food or water contaminated with fecal matter, kills a child every fifteen seconds, and in each decade that passes the number killed exceeds all World War II fatalities (George 2008; Curtis, Cairncross, and Yonli 2000). In India, forty-two children die each hour due to inadequate sanitation (Kar 2012; UN Millennium Project 2005; WaterAid 2007). Inadequate sanitation is one of the central drivers of disease outbreaks, including Ebola and Zika as well as Cholera, TB, and Typhoid. There is no area of investment that generates the extent of poverty reduction that sanitation brings (Black and Fawcett 2008; WHO 2012).

On the face of it, sanitation is simply the safe removal of human waste: a straightforward problem with a clear set of technological solutions. In practice, it is far from that simple. Sanitation is absolutely fundamental to everyday life in cities in the global South. Inadequate sanitation is intimately connected to inequalities in gender, caste, religion, education, and work. Yet, while the sanitation SDG aims to provide sanitation for all by 2030, almost half the countries in the world do not even recognise sanitation as a right, progress with provisions are often slow and patchwork, and global debates often underestimate the extent of the challenge or over-simplify solutions around technocratic interventions (GLAAS 2012).

Existing debates on sanitation in the global South tend to see sanitation as a *liberal* problem, and in two ways. First, the sanitation crisis is *seen* as a problem in particular kinds of ways. The profusion of global summits of practitioners, policymakers and academics, and the charts, numbers, stats and pronouncements, position sanitation as an absence of infrastructure (toilets, drains and water pipes). Second, the ensuing debates have focused predominantly on a series of discrete domains, and in particular appropriate and relevant technologies, finance and cost control, whether water should be considered an "economic good", whether to charge toilet "users" and by how much, whether good sanitation should be seen as the responsibility of the state, or the private sector, or the work of informal entrepreneurs or, increasingly, of "communities" (e.g. Kamal Kar's [2012] influential Community Led Total Sanitation). These debates are important and well-intentioned, and through them tangible improvements have been made to people's lives.

However, there are two closely related drawbacks to this liberal conception of the sanitation crisis. The first is that solutions often centre narrowly around technocratic issues such as getting the right materials, cost structure, or relationships in place (e.g. between governments and private companies or non-governmental organisations). The second is that the range of issues attached to sanitation – from race, gender and class, to land and housing, to education and livelihood – are often, while not in any sense forgotten, allowed to slip from view. The liberal conception of the sanitation crisis serves, in a curious way, to *sanitise* the crisis, and the lived nature of the urbanising sanitation crisis in the global South, its manifold politics and questions, its urban nature, and the potential ways forward, are profoundly truncated.

Rather than a liberal question, this chapter positions sanitation as a fundamental question of citizenship. Here, waste is not an abstract global humanitarian clarion call, but is a question of urban citizenship that is at once material and political, as much about things – pipes, drains, toilets, and so on – as it is about dignity, everyday life, and equality. When sanitation is placed in a context of urban citizenship, it is not only a question of governance and provisioning, but an urban question of bits of broken and inadequate materials, bodies and their waste, places and political economies, cultural politics and people, and an often profoundly contested politics of protest and resistance. What is revealed is less a liberal politics of managing sanitation supply and demand, and instead a material and deeply contested geography of coping with waste and making claims, as a fundamental politicisation of city provisions and city life.

What we start to then see is sanitation as a *networked problem*. Given that sanitation spills across multiple domains of life, habitation, and mobility in the city, it cannot be seen as an isolated issue. Sanitation bleeds into different spheres of urban life. It erodes health, deepens the exploitation of girls and women and limits their ability to move around the city, keeps children out of school, stops adults getting to work, exacerbates local tensions around religion or ethnicity, stunts bodily growth, curtails the nutritional value of food, becomes the vehicle of disease, becomes a fulcrum for urban protest and resistance, and so on. It prescribes people's ability to live fully in the city, to exercise what Henri Lefebvre once called their "right to the city".

In this sense, the arguments here resonate with the wider literature on networked infrastructure, and in particular on those debates focused on how urban life – in the absence of reliable, functional networks – is forced to improvise, adapt and customise all manner of informal and formal patterns of provisioning, whether in sanitation (Monstadt and Schramm 2017, water, energy (Silver 2015), transport (Fernando and Porter 2002), or other infrastructural forms. Stephen Graham and Simon Marvin's landmark book, *Splintering Urbanism* (2001), examined the "splintering" of public space and provisions in the context of urban infrastructure. They demonstrated how neoliberalism, and in particular the relations between privatisation, liberalisation, and the application of new technologies, shaped a globalising process of "unbundling" infrastructure. Their book precipitated an exponential increase in research exploring how infrastructure is either becoming de-networked, or, particularly in many poorer neighbourhoods in the global South, was either never networked or networked in quite different ways (Graham and McFarlane 2015; Coutard and Rutherford 2015). While the sanitation examples I draw upon are inevitably selective, taken together they resonate more widely with research on the politics of inadequate provisions in the city, from work on improvised energy provision, drainage, or water, to studies of inadequate housing, transport or health provisions (e.g. Amin 2014; De Boeck 2012, 2015; Satterthwaite and Mitlin 2014; Graham and McFarlane 2015; Lancione 2016; Ranganathan 2015; Shnitzler 2013; Silver 2014; Thieme 2017).

However, as Charlotte Lemanski (forthcoming) has argued, the relations between infrastructure and citizenship are often less well foregrounded in these accounts. Yet, the politics of sanitation I examine in this chapter demand a connection between infrastructure and citizenship. As we shall see, these are forms of politics that always operate by connecting material conditions to politics, connection toilets and other facilities to the obligations of the state, the national Constitution, rights, and gender equality. In her work on housing in Cape Town, Lemanski uses the term "infrastructural citizenship" to capture these kinds of relations. She argues that the connection between infrastructure and citizenship based on state programmes, denial and neglect, or in the contestation of the state, is not prominently made in accounts of the politics of infrastructure. Lemanski's definition of infrastructural citizenship is usefully broad, and so accommodates the diversity of ways in which, for example, activists make a claim to citizenship through infrastructure.

Given that many informal settlements are squeezed into small slithers of land and inhabitants are forced to live amongst human and solid waste, improving sanitation conditions often requires radical and significant political change. This includes, for instance, reform in land and housing rather than piecemeal attempts at toilet provision. In some cases, the challenges are less about infrastructure, housing or land than they are about tackling the racism or ethno-religious politics of local states. In short, understanding sanitation as a networked, multi-sided issue in the city means opening out an expansive imagination and debate about urban political transformation. The politicisation of urban sanitation is very often more than just a politics of toilet provision: instead, it is a politics of making claims to the city itself, of inequality and exploitation in the city, and of the right to a fuller city life. The ways in which those claims are made, however, varies considerably. In this chapter I examine two broad kinds: those focused on making claims in relation to local, contingent, unfolding struggles, and those focused on making claims to the wider city, for instance through the medium of rights.

What do "rights" mean in the context of sanitation and the city? More than just the right to decent toilets. Most of the time, addressing sanitation poverty and exclusion will require more than one sort of intervention – more than just technology, more than just political will, more than just money, more than just the right governance mix, more than just investment in adequate health care, more than just improved provisions in schools or public transport networks or in public places, and so on. In the absence of good sanitation across the city it becomes difficult to live a decent urban life. In other words, the right to sanitation is, in practice, *the right to city life*.

This is why movements like Mumbai's *Right to Pee*, which I discuss below, argue that their campaign is more than just for toilets. Instead, they see their struggle as simultaneously one of planning and culture, dignity and gender equality, and toilets and drains. For Henri Lefebvre, the right to the city was not just a right to housing or infrastructures or services, or to move freely through the city, as much as all these things matter. Instead, it was the right to be an active participant in the production of urban space. To be an active part of the production of urban space means making genuine contributions to urban planning (of neighbourhoods, infrastructures, services, schools, children's play areas, and so on), to what the city is and what it is for, to the material layout of the urban fabric. Right to Pee activists, as we'll see, adopt similar positions, and there are strong resonances with activists I briefly discuss in Cape Town, from the Social Justice Coalition.

The right to city life resonates with James Holston's (2008) description of "insurgent citizenship", in that it is a process of expanding citizenship rights by connecting infrastructure, bodies, informal neighbourhoods that are often deemed "illegal" by the state, and the participation in the life of the city. The ways in which the state fails to meet the rights of the citizen, such as the right to dignity, is quite different in both cases in the chapter, as are the ways in which activists and residents respond. However, the broader point is that there is a kind of infrastructural citizenship at work here. There is a long history of connecting sanitation and rights, from the action of civil rights activists and sanitation workers in post-war United States to historical struggles around caste and sanitation in India. Today, activists and residents in Mumbai, Cape Town and elsewhere are demanding the right to sanitation as the right to city life. In doing so, they take us beyond the abstractions of global human rights talk and into life in the city, its schools and colleges, clinics and hospitals and transport hubs, parks and public squares, homes and neighbourhoods, but also its political machine. Their successes to date may be partial, but they are vital illustrations of the power of rights-talk for mobilising a politics of sanitation in the city.

An urban question: Mumbai's Right to Pee

In 2011, CORO (Committee for Resource Organisations) for Literacy began exploring how best to respond to common challenges across its different affiliated groups in the Indian state of Maharashtra. Housing, malnutrition, violence against women, education, and other vital issues were discussed. One prominent member of the group in Mumbai – Mumtaz – suggested in one meeting that they might focus attention on toilets for women. The initial response was sceptical, some laughed, but Mumtaz put the case. Whenever there is a meeting or protest, she argued, there is rarely anywhere for women to go to use the toilet. If you move around villages or cities, there are often provisions for men – not enough, and often in poor condition, sure, but something nonetheless – and usually nothing at all for women. If there are toilets for women, they usually have to pay. Men, she argued, do not pay to use urinals most of the time, though in many public toilet blocks they will be charged for using the cubicles. Women, in contrast, are routinely charged by caretakers on the basis that caretakers cannot be sure whether a woman is using a cubicle for urinating or defecating. Women, Mumtaz argued, are being made to pay simply because they need to sit down and close a door behind them. The Constitution demands that everyone be treated with dignity and equality, so, she continued, isn't this a vital question not of toilet regulation but constitutional rights? In the end, people agreed to support the focus on rights, gender and toilets. A new movement was born that has since gathered local and global attention: Right to Pee.

CORO's Mumbai group has historically operated in Chembur, just south of M-East. When the Right to Pee (R2P) movement was formed– an amalgamation of CORO and 32 other civil society groups in the city promoting change on a set of connected themes (gender, sanitation, water, housing, land, and others) – the agreed first step was to develop a better informed picture of sanitation in the city. Data was what was needed, data that could hold the city and state authorities to account. They began by surveying 129 city toilet blocks. This data was then set against the BMC's (BrihanMumbai Munipcal Corporation's) own data, made available through a Right to Information request. The request revealed that the BMC was making public promises on sanitation improvements with no discernible budgetary increases, and that while the BMC's data accounts for 10,000 toilet seats, 63% of which are exclusively for men, the provision for women was effectively zero (Table 3.1). "They were just playing with statistics", said one activist. R2P activists extracted promises from the BMC but nothing happened. Local media initially did not pick up the campaign, with the exception of one Marathi newspaper, but a piece in the *New York Times* in 2012 (Yardley 2012) led to a spike in media reporting which put further pressure on the BMC. "I won't say the media attention was absolutely essential" for subsequent BMC decisions, reflected Supriya Sonar, a R2P activist, "but it played a very important part".

A year later, in 2012, the activists launched a signature campaign on the railway stations. The domestic railway in Mumbai is the busiest of any city in the world, used by an estimated 7.5 million people per day. The signatures called for an improvement in provisions beyond the woefully inadequate and usually broken or poorly functioning fragments that exist. From the start, then, the campaign has not only been about low-income neighbourhoods, but about sanitation experiences *across the city* – in town, in transit, near home, and for everyone from low-income vegetable vendors selling material in the city to middle-class commuters. Supriya has said: "Today, the Right to Pee is everyone's campaign – from women fruit vendors to doctors and educationists, to town planners and gender experts" (cited in Patel 2013: no page). Another activist in the movement, Usha Kale, has put it this way: "Ours is a movement of sewage cleaners and sweepers, flower sellers and fishwives. Just the sort of women who are used to a fight" (Faleiro 2014: no page).

But for all the importance of these connections and solidarities, there is also an important class differential to these gendered politics, and here class can be mobilised to work not just for but *against* the expansive vision of inclusion and provision that R2P campaigns for. For example, when R2P activists stood on railway platforms soliciting signatures in support of the movement, it was noted that some women ignored the petition or even on occasion "stopped only to underscore their privilege. 'Who needs public toilets when you have toilets in malls?'" (Faleiro 2014: no page), one woman remarked. If there is a cultural politics of "pollution" inscribed upon the female body in India (Phadke et al. 2011) – linked to menstruation, ovulation, lactation, and notions of contamination – those politics do not take shape in the same sorts of ways for all women and are entrenched by class (Shah 2014; Bapat and Agarwal 2003; Peake 2015).

Table 3.1 From Mumbai activist organization Right to Pee

Women population as per 2011 census	*5,741,632*
Total no. of "pay and use" toilet blocks	2,591
Total no. of toilet seats	For men: 6,568 For women: 3,813
Total no. of bathrooms	842
Total no. of urinals	For men: 2,849 For women: 0 (none)

Source: Supriya Sonar, R2P

Alert not just to the need for sanitation rights but to *rights in relation to social difference*, R2P pressed the BMC to provide facilities that have social vectors factored in from the start, which are built for instance with children, the disabled, the blind, and the elderly firmly in mind. The movement is planning to move towards a system of online mapping of toilets, where residents can upload images of problems in a real-time database, but at the moment there isn't the technical expertise to make this a reality (they are also in touch with activists in Hyderabad who mapped community toilets). R2P has put out for local "right to pee journalists" (Figure 3.1), residents who report problems locally, mostly older residents who are frustrated with living with fragments of sanitation.

In 2013, state women and child development minister Gaikwad was petitioned by R2P to provide public toilets for women free of charge. R2P delivered a list of 50,000 names, gathered mainly from the city's railway platforms, partly because the BMC had been slow to respond to demands for better provision for women in the city. Gaikwad helped introduce the *Maharashtra Policy for Women* in 2013, which mandates the construction of a women's toilet block every 20 kms (12.4 miles). The fact that just one block every 20km constituted an improvement in provisions is itself a powerful reflection of just how fragmented sanitation provisions are for women and girls, and poor women and girls in particular, around the city. This move in turn pressured the BMC to act with more haste to build female toilets, although there has been little substantial construction on the back of the flurry of public statements. Moreover, the movement continues to encounter obstacles or dismissive responses from state and city officials at different levels. One state official, for example, while conceding that R2P had some good ideas such as providing toilets for women along highways, argued: "They make a lot of noise, [but] they are very small organization", insisting that most of their proposals were

PRESS RELEASE FOR TODAY

"Be a Right to Pee Journalist and share "Life Around Toilet" in your community
Main bhi journalist"

"Right to Pee" Team hereby announces "Main bhi journalist" (Right to Pee Journalist) for clean, safe, free public urinals for women. The RTP team has been advocating with the Mumbai Municipal Corporation for the last four years for the provision of this basic amenity in the city. All the citizens can be Right to Pee Journalist

- Those who believe in change
- Those who want their own community, city to be safe and clean
- Those who feel Citizens rights to be respected

Those who want to be part of "Main bhi journalist" pls call us between to 3pm to 5pm from 13th Sept. to 23rd to Sept. 2015.

Figure 3.1 Right to Pee's call for "pee journalists"

actually already being addressed by the BMC, and that the claim that women had to pay was exaggerated.

Alongside collecting data, lobbying the BMC and regional state, working with municipal architects, and monitoring local conditions, other tactics have been used. Mahila Milan ("Women Together", community-based groups) activists linked to R2P have on some occasions spent days sitting in front of their local ward offices, demanding that staff be suspended if demands weren't met. There have been protests at the BMC office in town. At one point, frustrated by the refusal of officials from the Government of Maharashtra to meet with them, activists considered a "pee protest" outside the main state office in south Mumbai. The Chief Minister had refused to meet activists from R2P, and after attempts over a couple of years the idea for the pee protest came from Mahila Milan activists and gathered a lot of attention across the R2P network. The idea circulated social media and was covered by the *Asian Age* newspaper. The next day they got an appointment with senior state officials; the threat of the pee protest was withdrawn.

If a gendered urbanism of sanitation exclusion is to be tackled, then writing change into the planning process itself is one vital step. As Mumbai's latest development plan progressed through its various drafts, R2P have been arguing that toilets need to be identified as public amenities in the plan of public places (schools, markets, transport stations). The first draft of the 280-page report did not even mention sanitation, but the BMC has been increasingly receptive, and by May 2016 the revised draft had more to say about sanitation, for example suggesting that improving provision could result from adding an extra story to some 5000 existing public and community toilet blocks (Sarkar 2016). This entails rebuilding some structures that are already weak, and a budget has been allocated as part of the proposal. R2P has been working with the Architecture department in the BMC to design new city toilet blocks, blocks where there are separate male and female caretakers, which are wheelchair friendly, and which are easier to clean. Architects in the department explained that they had little experience collaborating with activists, and no experience designing toilet blocks, and suggested that the very fact this was new reflected a prioritization of sanitation by the municipality. Activists and officials have gone back and forth on several designs – for example, while some of the architects had easy to clean materials like aluminium and acrylic in the initial designs, R2P activists argued that these would be more expensive and perhaps not robust enough in areas with heavy use.

Each year, as Supriya at R2P pointed out, the BMC has increased provisions for sanitation in its budgets, and the legal basis for enforcing rights has strengthened. For example, in late 2015, the Mumbai High Court responded to a Public Interest Litigation filed by activist groups around health and the rights of women in relation to public toilets. The petitioners invoked Article 226 of the Constitution, which empowers High Courts to interpret the Constitution for its own territories and argued that the nonexistence or poor condition of public toilets led to safety concerns, non-accessibility or unhygienic conditions, and medical concerns because women were forced to hold in urine for extended periods of time. The High Court agreed, and consequently insisted that toilets must have trained maintenance staff and reasonable charges, and concluded: "[Women] need these facilities at public places like Railway Stations, Bus Stands, Banks, Public Offices like State Government Offices/Municipal Offices. Public health is of paramount importance and it is the duty of the State and the Corporations to ensure that public latrines, urinals and similar conveniences are constructed, maintained and kept in a hygienic condition. The need for toilets is felt even more acutely, during menstruation" (Maharaj 2016: no page).

These are important and welcome words, but there is a long way to go. On 8 March 2016, on the occasion of International Women's Day, R2P returned an award that the city's Mayor, Snehal Ambeka, had presented to them for their work on sanitation and gender. Supriya told the *Times of India*: "Nothing has changed on the ground. The BMC budget has no provisions for women's toilets in the city and focuses only on household and community toilets. Our demand too falls under the Swachh Bharat [Clean India] campaign" (Pinto 2016: no pagination). Some state officials, unsurprisingly, see things differently. One senior official in Mumbai, insisting that the municipality is a "massive machinery and financial power which is unparalleled to any other city in India", was dismissive of R2P and the claims they were making. She argued that the BMC, through the Slum Sanitation Programme (SSP), had provided 80,203 toilet seats to "slums" over a 20-year period (Sharma and Bhide 2005; McFarlane 2008a, 2008b), improved maintenance and increased awareness of the importance of sanitation amongst "slum dwellers". For example, she claimed that through awareness-raising campaigns – from information trucks with PA systems and street plays focusing on hygiene, to information disseminated through the media and social media – the BMC had reduced open defecation rates from 58,000 families (in the 2011 census) to 15,000 families in 2015. She added that 117 "open defecation hotspots" had been located with the aim of eradicating OD by 2017, adding that Wards B and C were already "OD free". Not surprisingly, many activists find claims like this preposterous, and see them as politically motivated attempts to defend and promote the state that mask the realities of urban sanitation rather than face them.

R2P activities involve the intertwining of different groups in a changing assemblage of like-minded and relevant people, including those who have different agendas, from toilet block operators to local BMC officials who often accompany R2P activists on their inspections. This loose and sometimes contradictory assemblage reflects R2P's position of collaborating with the BMC where possible, and confronting where not. We see this, for instance, in the ward-level activists who help monitor local toilet blocks. Hemanth, for example, is from Govandi in M-East and worked for an NGO that became linked to CORO and, subsequently, R2P. He now coordinates some of R2P's work in M-East, organising meetings every couple of months with interested residents, initiating new members and training them in the checklist process for inspecting community toilet blocks with the local BMC Junior Officer. Or take Maruti, an auto-rickshaw driver from a low-income neighbourhood in Ghatkopar. Maruti has long been interested in the challenges facing drivers, and took these issues up as a CORO Quest Fellow in 2013–14. This is when he encountered R2P. He now does much the same work Hemanth does, but in N-ward. Here, he says, women often have to wait two or three hours in queues to use toilets.

These R2P volunteers and BMC local officials use checklists to score toilet blocks from "A" to "D". A toilet block rated "D" is blacklisted and, eventually, the group running the block may have their licence removed by the BMC. The checklists are tools for evaluating the conditions of provisions, for trying to ensure that they are as functional as they can be, and for mitigating the archipelago of sanitation provisioning. At the same time, the checklist can open out new ways of working with sanitation infrastructure, and not always for the best. For example, the blacklisting system, one R2P activist claimed, has its drawbacks in that some BMC officials have used the threat of it to raise bribes.

The collaboration with the BMC has meant that Hemanth, Maruti, and others have added authority when inspecting the blocks, even if, some activists suggested, most BMC officials see R2P activists as very much the juniors in the collaboration. This relationship is exacerbated by the fact that some toilet block operators are run by BMC officials, making money as a side operation. Hemanth and Maruti pointed out that initially toilet block operators were sceptical of R2P, and sometimes confrontational. Maruti pointed out that after one public meeting he received a phone call from a powerful individual who owns several blocks, warning him not to create problems. Sahar, another volunteer, has had similar experiences, and one operator even complained to a local politician who then in turn confronted her. A great deal of the work of inspection has been explaining and reassuring those who operate toilet blocks that R2P's only interest is to improve conditions and to do so by working with them rather than against them, including through improving working conditions for caretakers and cleaning staff.

This inspection work is often unpredictable, especially in cases where the operator has contracted out operations to a group of local people. Sometimes the inspections have unintended consequences in that the caretaker may be fired but conditions themselves do not improve. The social configuration shifts, but the material situation doesn't. One of the challenges has been holding caretakers to account while recognizing that caretakers are themselves often marginalised, lower-caste residents. For Supriya, caretakers need to be on board, and are essential for change to occur. On many toilet blocks now where R2P work there is a large display board stating that "urinals for women are free", which at least brings a public clarity for users and caretakers.

Other operators treat the blocks as commercial enterprises rather than local services, and refuse to take on anything that increases costs. Operators often overcharge (R3 per use, but sometimes costs are found to be as much as R10 per use), and some of the profit is sometimes used to pay off well-placed local connections to spin a positive story to inspectors. Often residents are unaware of what they should be paying, or even if they should be paying at all (some blocks are supposed to be free of charge), or unaware if they are paying to use the block or for its maintenance. Some blocks improve and then return to old habits, becoming ever less well maintained, underlining the need for repeated inspections. But increasingly, activists say, toilet block operators come to local meetings with residents and show a willingness to work with R2P. Throughout, volunteers like Hemanth, Maruti and Sahar exchange experiences, often through Whatsapp, sharing what works or asking advice on particular issues as they come up.

More generally, CORO's network of Mahila Milan (Women Together) groups have been surveying the conditions of toilet blocks, the activities of toilet block caretakers, and sharing experiences. Each of these activists – over 20 of them, mostly women – take the checklist to toilet blocks and survey conditions, and as a result 96 areas have been identified in need of improvements either to the physical conditions or the ways in which women are treated by caretakers. The BMC, in meetings, has told the activists it will follow up on these, but that it will take longer on land that is privately owned. The most common retort from the BMC is that there isn't enough space to build toilet blocks in dense areas, an argument R2P understands but rejects as an excuse for inaction.

The inspection activities do vital work in improving broken toilets, maintenance, and services to toilets (water, electricity, drainage etc.), and they illustrate the different ways in which provisions connect the social and material in specific expressions. Local power relations, structures of ownership, forms of intimidation and violence, collaboration and solidarity, and social relations of caste, gender, and class, all play a role in shaping the ways in which sanitation takes shape. When the inspection work is able to go well, sanitation provisions can become more reliable. But this is a laborious, slow, and changing process, as much social, economic and political as it is material and bodily.

R2P's work entails shifting the registers through which sanitation provisions are constituted as political. On the one hand, sanitation poverty and exclusion are expressed as an affront to gender, dignity and constitutional rights, and in negotiations with local and regional state, are generative of substantive questions of *infrastructural citizenship* (Lemanski, forthcoming). As Sonia Faleiro (2014: no page) writes of R2P, "the unprecedented acknowledgement of a woman's right to a public toilet [at senior policy levels] was seen as a victory not just for the fight for better sanitation, but for the women's movement." Indeed, R2P's Supriya argues that the campaign has "evolved" from "right to pee to right to city". In the neighbourhoods in which CORO operates, most people talk about the "mutari" (urinals) movement rather than "Right to Pee" – Right to Pee is a slogan that resonates in the English media and across different publics in the city – but the use of "rights" is of course no happenstance of language. Mumtaz remarked: "I'm looking for my place in the city. My own safe place, with dignity".

The question of toilets is for her a question of equality, of protecting women from violence and promoting empowerment, and of actively producing and being part of the planning and life of the city. Supriya argued that "it's discrimination based on gender – it's not just that there is a charge [to use toilets]…it's not about facilities, it's a political statement", while Sujata talked about the "freedom" to participate in the city, to move around and not be stuck indoors, and repeatedly asserted that the struggle was about "citizenship". If there are no provisions, Sujata argued, then the city is saying to people: "Shut up and stay home". Instead, added Supriya, R2P is asking: "How do you claim your city as a citizen?"

The fact that facilities are absent or inadequate, then, is not just a question of material provisioning, but of political citizenship. Citizenship here is understood as a question of equal rights to move through the city, for work, pleasure or other reasons, without fear of being unable to get to a sanitation facility should it be needed. At the same time, while activists here insist that the struggle is about more than just one of material provisioning, the politics, economies, and socialities of material toilets are clearly fundamental.

Given this, R2P's work also entails a more grounded, immersive and responsive approach that seeks to find whatever openings there might be just to get at least some things improved or protected at local toilet blocks, working with and making claims on local power-brokers. This is what makes R2P such an interesting movement: it moves continually between distinct politics of claim-making, one focused on the local and contingent, and the other focused on the wider rights of women in the city. R2P's different instantiations of the political, then, reveal a citizenship struggle that is at once material and, activists insist, much more than just material. It is about things *and* equality, about objects *and* the right to the city, about material conditions *and* the Indian Constitution. The struggle for citizenship, then, is at once material and political, a set of claims posed through the medium of infrastructural citizenship.

Against this background, sanitation emerges as a kind of networked concept in a context where the infrastructural network does not exist in a reliable sense (Coutard and Rutherford 2015). As a networked concept, sanitation here ranges through toilets and pipes and maintenance of often barely functioning fragments to data, accountability, and rights. To see sanitation as an urban question is to see that, from poor neighbourhoods to public toilet blocks, railway stations and parks, the politics of urban sanitation is expressed in multiple contexts that often exceed the provisions themselves. What R2P offer here is not just a recasting of how we might understand sanitation in the urban context, but a way of *seeing like a city* (Amin and Thrift 2017; Kishik 2015; Magnusson 2011), a form of seeing that seeks to dismantle discourses of progress by exposing a city fragmented by politics and culture while attempting to generate new possibilities, and which works amid grounded and often messy particularities and on conditions in the city as a whole. In the next section, I explore a similar movement in Cape Town.

Accounting for the city

The Social Justice Coalition (SJC) is a high-profile movement in the city that emphasises the process of auditing urban conditions for political intervention in sanitation, using the language of the state to hold it to account through a monitoring of toilet conditions in the large township of Khayelitsha: "Our objective is to make government accountable", one activist said. Like R2P in Mumbai, the SJC has a history of campaigning for rights in the context of often violent xenophobia. As with the surveys R2P are engaged in, the SJC tactic of auditing is in part about seeing like a state (Scott 1998), and is a form of politics that entails, in Steven Robins' (2014: 107) description, the more "mundane technical and bureaucratic work of making the state responsive to the needs of the urban poor." SJC employ a politics of critically engaging with the municipality on the terrain of data, standards, and measurement, the city engineer, the statistical, and the ordinary geography of maintenance, operation and repair. In speaking such a language, SJC maintains a distance from political parties: "We need to see beyond boundaries of politics", one SJC activist explained, "we want to talk about people, rights and life". SJC (2014: 8) has connected rights to both the Constitution and the municipality, and called for a shift from ad hoc provision to regular and predictable infrastructures and services reflected in a proper urban plan:

> According to the Constitution, local government is responsible for progressively realising the right to basic sanitation. The failure to do so violates the right to human dignity, freedom and security, particularly for people living in informal settlements...The City of Cape Town does not have a detailed, integrated, time-bound plan in place to progressively realise the right to basic sanitation. (McFarlane and Silver 2016)

SJC was formed in 2008 as a response to xenophobic violence mainly against southern African migrant communities across South Africa, including Cape Town, that left dozens dead and thousands displaced. It is a mass member organisation working with Ndifuna Ukwazi ("Dare to Know"), a technical support NGO. SJC rose to prominence in the sanitation debate through the auditing of toilets in their 2010 "Safe and Clean Sanitation" campaign (see Figure 3.1), involving the mapping and auditing of sanitation provisions in Khayelitsha and the publication of data through reports detailing the often insanitary conditions and long delays in repair and maintenance faced by residents (SJC 2014). For instance, the 2014 audit undertaken in parts of Khayelitsha found that 49% of toilets were "dirty" or "very dirty", with one in four flush toilets not working and 50% of interviewed residents reporting that janitors never cleaned the toilets in their area (SJC 2014). Much like the surveys and checklists of R2P, these audits present a snapshot of an urbanism of inadequate sanitation, with the aim that the data might both hold the state to account for its data, and be levered into negotiations with the state.

In early 2015 the auditing was further developed through the piloting of real time data monitoring and reporting by Ndifuna. This allows residents, using their smart phones to upload images and text onto a database, to instantly communicate faults to the municipality. This is an important advance on the current reporting system set up by the CCT (City of Cape Town), which requires having phone credit or walking to the few public phones in Khayelitsha and often paying for an expensive call while kept on hold. The SJC real-time system is about "bringing [the] city to the people's doorstep, [we] want them to have access to report issues without using their own money". Here is an attempt to make claims around sanitation in quite a different way, by creating an opportunity for residents to connect provisions, data, and a larger rights-based campaign. Like R2P, SJC see sanitation as a rights-based struggle for citizen provisions, and the movements mirror one another in making the shift from fragments – the inadequate provision and maintenance – to the whole city, a questioning of urban inequality and the role of the state. For example, one of the contested issues that the auditing has examined is the different actors involved in maintenance. This has consisted of two elements.

First, the provision of chemical toilets has been contracted out to a private company, Mshengu Services. This privatization of public services in a R140 million contract has become fraught and subject to regular complaints about insufficient conditions and maintenance, and one SJC activist accused the CCT of "running away from accountability". The SJC auditing of Mshengu Services, particularly in relation to maintenance of the toilets, led to an investigation by the South African Human Rights Commission. At the same time, echoing R2P's emphasis in the north-east of Mumbai, SJC retains its place-specific focus on the politics of sanitation provisions, especially in Khayeltisha where the movement is based and meets. Second, and following on, the auditing has raised questions about the level and flow of money. For example, the fact that janitors are funded through a national scheme – the Expanded Public Works Program (EPWP) – rather than the CCT itself, and on short-term low-wage contracts at that, has fed debate around CCT's commitment. SJC has been careful to lay blame for poor maintenance not at the feet of exploited janitors but at the CCT – as one janitor complained: "CCT has the money." These two processes underline the need for the kinds of data that the SJC is collecting, and remind us that collecting data always impinges on the political terrain of what the post-apartheid state does and does not deliver its citizens. In other words, collecting data and using it to raise questions can effectively connect local fragmented provisions to a politics of the urban whole, here in the form of rights that connect city and national Constitution.

With the aim of ensuring that the municipality delivers the socio-economic rights established in the Constitution and meets its own obligations to infrastructural citizenship, SJC produces reports detailing sanitation conditions or reactions to municipal budgets. The municipality, as Parnell and Pieterse (2010) argue, needs to be able to subsidize and create adequate institutions for the provisions of daily life, and it needs to have the capacity and data as well as the will and the money to deliver infrastructure at scale. The data-based claims of movements like SJC are vital in this context, because they push the state to explain why an urbanism of sanitation fragments persists across the city, and how it can or cannot meet basic provisions like sanitation. In doing so, a contested but crucial discourse on the whole, and its various relations to the messy worlds of urban sanitation, is positioned more centrally on the urban agenda. While the "sanitised" charts and data used by international agencies mentioned in the opening pages are often detached from these messy worlds of urban sanitation, R2P and SJC reveal the political potential of a more grounded, less polished version of similar kinds of data, this time created by activists and residents, by showing how it can become essential in bringing the state to accountability.

However, the process of holding the municipality to account has been deeply politicised, met with negative reactions from the local state and very public disagreements (especially across social media) about the way in which the auditing is produced, disseminated and used. Of particular frustration to the municipality has been the perceived holding of data on faulty toilets to be published months after collection in the audit reports, rather than instantly informing the appropriate department, in order to increase the impact of data and the likelihood of generating political attention and debate more generally in the city. Despite the hostility between the CCT and SJC, the objective of the auditing work remains focused on addressing urban sanitation by creating a working relationship. As an activist from Ndifuna reflected, following a visit to meet Indian organisations undertaking similar work: "We will get there. One of the things I learnt in India is that government was once against social audit process, at some point they understood NGOs are only doing this to work with government and make sure spending goes on what it should".

As with Mumbai, the discourse in Cape Town surely needs to shift towards one of more sustained distribution of land and resources rather than a form of divide and rule predicated on occupation, temporality and the migrant figure (Parnell and Pieterse 2010). SJC signals one route forward here, one that works with inadequate provisions *and* rights by holding onto social and spatial difference, but insisting on the centrality of equality for everyone across the city from home to work and in transit. In quite different ways from the Mumbai example, then, the SJC activists develop a form of infrastructural citizenship that connects data, materials, and the state, an infrastructural citizenship that is set towards – as with R2P – not just the right to the city, but the right to city life.

Both cases are active forms of citizenship that are at once material and political, and which insist that sanitation has always been and must continue to be understood as both. Part of the power of both movements is that they place claims in ways that very directly play on the historical languages and processes of the state and its obligations to the Constitution and citizenship, for instance by using data, and through an appeal to equality. The fact that this play of data, materials, equality, and historical obligation are forged in relation to a fundamental provision – sanitation – means that the struggle for infrastructural citizenship in these two contexts is one that the state cannot simply ignore.

Conclusion

The politics of sanitation in the city is both singular and multiple, and the claims to improve urban sanitation provisions are made in quite different ways, of which this chapter has examined just two broad forms. Examining the ways in which provisions fall apart, are put together, maintained, experienced and politicised reveals poverty and infrastructure to be relations extending beyond income and materials to gender, religion, caste, class, party politics, and different orders of political expression (Satterthwaite and Mitlin 2014; Elwood et al. 2016; Lawson 2012; Moser 2009; Banks 2015). The examples explored here reveal quite distinct forms of infrastructural citizenship in how they make claims on the state that are simultaneously about material provisioning and rights in and to the city (Lemanski, forthcoming). In doing so, they raise important questions about how a politics of infrastructural citizenship can both operate in place and at scale in the city. Given that the life of urban sanitation is profoundly geographical and temporal (Parnell and Pieterse 2014; McFarlane et al. 2014), and that a politics of rights can entail "a conceptual disentanglement of things from their actual sociomaterial contexts" (Björkman 2015: 231), how might we keep a critical hold of both local struggles and the struggles for rights to sanitation across the city as a whole?

In *Reducing Urban Poverty in the Global South*, David Satterthwaite and Diana Mitlin (2014) argue that for all the strength and importance of rights-based approaches, they can be insensitive to local context, can undermine commitment to incremental processes of working with the sorts of provisions I've considered in this chapter, ignore the necessary trade-offs between issues and power-brokers which inevitably characterise urban change and development, and risk reproducing the divide between the formal and the informal in that it is the "formal" – more easily "seen" by the state – that is more likely to benefit from provisions based on rights. Residents often don't have the time for or interest in struggles for abstract rights, and instead if they seek out civil society groups or social movements at all, it is often because there is a specific issue that might be practically addressed. Campaigns for rights sometimes move on past the urban poor themselves, and middle- and upper-income groups are often in a better position to take these struggles forward, given that "they know the law, their housing is legal and most work within the formal economy" (Satterthwaite and Mitlin 2014: 43). Even where there is a measure of political will to provide, there are no guarantees that solutions will work out well for people. For example, in her research on water and sanitation in Delhi, Yaffa Truelove (2011) has shown how providing "legalised" water pipes can exclude poorer women due to the quality and peripheral location of the water.

Satterthwaite and Mitlin do not reject a politics of rights and democratic citizenship, but ask us to critically think through how rights are framed and what their effects might be, both in and of themselves and in relation to the imaginaries of the city that tend to accompany them (for example, an often gendered imaginary of formal and informal, legal and illegal, state and citizen; see Peake 1993, 2015). A politics of rights can construct a quite particular imaginary of the whole city, of the nature of the political, and of struggle. It is a quite different form of politics from the messy complexity of the city with which Mumtaz, Hemanth, Maruti and others are routinely embroiled, with its multiple and changing authorities, and which requires closer attention to the "socionatural processes and politics of urbanization and contestation through which the actually existing city is being made" (Björkman 2015: 233).

R2P attempt to steer between two forms of politics, both of which seek to address an inherited urbanism of fragmented sanitation provisions: one around rights and plans and another around the heterogeneity of urban authorities and knowledges. There are risks and limits in making this kind of move, but their work is a useful guide for activists (and policymakers or urban researchers) looking to try to find ways of negotiating between and keeping a hold of both a politics of local provisions and rights to the city. To work with both the heterogeneity of multiple interests and forms of power that can enter into the everyday assembly and disassembly of urban provisions, *and* demand rights as a politics that would address urban sanitation across the whole city, is a powerful position from which to see like a city (Amin and Thrift 2017; Magnusson 2011). R2P provide us with one possible route-map through which to maintain a grasp of the specificity of provisions and the different ways in which they are inhabited within (and between) cities, while also keeping a hold on some of the drivers of fragmentation.

There are instructive lessons here for how critical urban researchers might reflect on the place of rights in progressive urban change. A recurring theme in Henri Lefebvre's *The Production of Space* (1991[1974]) is that of the analytical and political potential of moving between localities and the wider politics of the city. He mentions in passing, for instance, that concerns over pollution, and the environment more generally, can serve to shift political attention from a specific issue to the whole in the form of the city or the planet. The key question, for Lefebvre (1991[1974]: 326), is "which kind of whole and whose and with what consequences for whom?" For Elizabeth Grosz (2005), writing not about cities or space but about bodies, one productive route is to focus on a politics of *becoming*. Becoming is both in part about recognition (e.g. rights), but is necessarily more than that; becoming is always a generative process. The politics of becoming embraces rights but insists that urban living and possibility must necessarily exceed the focus on rights per se.

Grosz finds in Deleuze's thought not a rejection of rights and identity politics – Deleuze and Guattari recognise in *A Thousand Plateaus*, for instance, that such a politics remains vital – but an insistent pressure to push beyond rights to break into new ways of being and relating that do not fix and freeze categories like "woman" (Deleuze 2001[1968]). If left to a politics of rights and recognition alone, the category of woman – or city – can become a knot that both strengthens and fixes, both enables and closes off possibilities. As

Rachel Colls (2012: 435) argues, the potential of Grosz's understanding of the female body is that bodies "are not understood as the product of a patriarchal culture but are excessive to hierarchical control". The body is more than its visible body-image, more than the discursive, geographical and political limits placed on it. This is useful for thinking of a politics of local struggles and rights together, a politics not just of seeing the margins and the absence of rights, but seeing the contents and differences in the ways that bodies can and do become expressed politically (Braidotti, 2003; Irigaray, 2004; Kristeva, 1982). For R2P and SJC, the becoming of infrastructural citizenship can never be about rights alone, just as efforts to work with and improve local provisions that require often laborious and changing place-specific negotiations cannot be the end point of efforts to address the fragmented urbanism of sanitation.

Acknowledgements

Thank you to Charlotte Lemanski for editorial guidance and constructive feedback. A big thank you to Ankit Kumar and Jonathan Silver, both of whom worked with me on some of the research related to this chapter. The fieldwork reported on here was supported by a Leverhulme Prize.

References

Amin, A. (2014) Lively Infrastructure. *Theory, Culture and Society.* 31(7–8): 137–161.

Amin, A. and Thrift, N. (2017) *Seeing Like a City.* Cambridge: Polity Press.

Banks, N. (2015) Livelihood limitations: The political economy of urban poverty in Dhaka, Bangladesh. *Development and Change.* 47(2): 266–292.

Bapat, M. and Agarwal, I. (2003) Our needs, our priorities: Women and men from the slums in Mumbai and Pune talk about their needs for water and sanitation. *Environment and Urbanization.* 15(2): 71–86.

Björkman, L. (2015) *Pipe Politics, Contested Waters: Embedded Infrastructures of Millennial Mumbai.* Durham, NC: Duke University Press.

Black, M. and Fawcett, B. (2008) *The Last Taboo: Opening the Door on the Global Sanitation Crisis.* London: Earthscan.

Braidotti, R. (2003) Becoming woman: Or sexual difference revisited. *Theory, Culture and Society.* 20: 43–64

Colls, R. (2012) Feminism, bodily difference and non-representational geographies. *Transactions of the Institute of British Geographers.* 37: 430–445.

Coutard, O. and Rutherford, J. (Eds.) (2015) *Beyond the Networked City: Infrastructure Reconfigurations and Urban Change in the North and South.* London: Routledge.

Curtis, V., Cairncross, S. and Yonli, R. (2000) Domestic hygiene and diarrhoea—Pinpointing the problem. *Tropical Medicine and International Health.* 5(1): 22–32.

De Boeck, F. (2012) Spectral Kinshasa: Building the city through an architecture of words. In T. Edensor and M. Jayne (Eds.), *Urban Theory Beyond the West: A World of Cities.* London: Routledge.

De Boeck, F. (2015) Divining the city: Rhythm, amalgamation and knotting as forms of urbanity. *Social Dynamics,* (ahead of print): 1–12.

Deleuze, G. (2001 [1968]) *Difference and Repetition.* London: Continuum.

Elwood, S., Lawson, V. and Sheppard, E. (2016) Geographical relational poverty studies. *Progress in Human Geography*. doi: doi:10.1177/0309132516659706

Faleiro, S. (2014) For some voters in Mumbai, this election's all about toilets. *Quartz*, 1. http://qz.com/196893/for-some-voters-in-mumbai-this-elections-all-about-toilets/ (accessed 10 May 2016).

Fernando, P. and Porter, G. (Eds.) (2002) *Balancing the Load: Women, Gender and Transport*. London: Zed Books.

George, R. (2008) The Big Necessity: Adventures in the World of Human Waste. London: Portobello Books.

GLASS (2012) *UN-Water Global Analysis and Assessment of Sanitation and Drinking-Water: The Challenge of Extending and Sustaining Services*. Switzerland: World Health Organization.

Graham, S. and Marvin, S. (2001) *Splintering Urbanism: Networked Infrastructures, Technological Mobilities and the Urban Condition*. Oxford: Blackwell.

Graham, S. and McFarlane, C. (Eds.) (2015) *Infrastructural Lives: Urban Infrastructure in Context*. London: Routledge-Earthscan.

Grosz, E. (2005) *Time Travels: Feminism, Nature, Power*. Crows Nest: Allen and Unwin.

Holston, J. (2008) *Insurgent Citizenship*. Princeton: Princeton University Press.

Irigaray, L. (2004) *An Ethics of Sexual Difference* (C. Burke and G.C. Gill, trans). London: Continuum.

Kar, K. (2012) Why not basics for all? Scopes and challenges for community-led total sanitation. *IDS Bulletin*. 43(2): 93–96.

Kishik, D. (2015) *The Manhattan Project: A Theory of the City*. Stanford: Stanford University Press.

Kristeva, J. (1982) *Powers of Horror: An Essay on Abjection* (L.S. Roudiez, trans). New York: Columbia University Press.

Lancione, M. (ed.) (2016) *Rethinking Life at the Margins: The Assemblage of Contexts, Subjects and Politics*. London, New York: Routledge.

Lawson, V. (2012) Decentring poverty studies: Middle class alliances and the social construction of poverty. *Singapore Journal of Tropical Geography*. 33(1): 1–19.

Lefebvre, H. (1991[1974]) *The Production of Space*. Oxford: Blackwell.

Lemanski, C. (forthcoming) Infrastructural citizenship: Spaces of living in Cape Town, South Africa. In K. Ward, A.E.G. Jonas, B. Miller and D. Wilson (Eds.), *The Routledge Handbook on Spaces of Urban Politics*. pp. 350–360. doi: doi:10.4324/9781315712468

Magnusson, W. (2011) *Politics of Urbanism: Seeing like a City*. New York: Routledge.

Maharaj, K. (2016) Bombay High Court makes Right to Clean Toilets a Fundamental Right for Women in India. Oxford Human Rights Hub, 8 February, http://ohrh.law.ox.ac.uk/bombay-high-court-makes-right-to-clean-toilets-a-fundamental-right-for-women-in-india/ (last accessed 29 July 2016).

Mara, D. (2012) Sanitation: What's the real problem? *IDS Bulletin*. 43(2): 86–92.

McFarlane, C. (2008a) Sanitation in Mumbai's informal settlements: State, 'slum' and infrastructure. *Environment and Planning A*. 40(1): 88–107.

McFarlane, C. (2008b) Governing the contaminated city: Sanitation in colonial and post-colonial Bombay. *International Journal of Urban and Regional Research*. 32: 415–435.

McFarlane, C., Desai, R. and Graham, S. (2014) Informal urban sanitation: Everyday life, comparison and poverty. *Annals of the Association of American Geographers*. 104(5): 989–1011.

McFarlane, C. and Silver, J. (2016) The poolitical city: 'Seeing sanitation' and making the urban political in Cape Town. *Antipode.* 49(1): 125–148.

Mitlin, D. and Satterthwaite, D. (2013) *Urban Poverty in the Global South: Scale and Nature.* London: Routledge.

Moser, C. (2009) *Ordinary Families, Extraordinary Lives: Assets and Poverty Reduction in Guayaquil, 1978-2004.* Washington, DC: Brookings Institute Press.

Parnell, S. and Pieterse, E.A. (2010) The 'right to the city': institutional imperatives of a developmental state. *International Journal of Urban and Regional Research.* 34(1): 146–162.

Parnell, S. and Pieterse, E.A. (eds) (2014) *Africa's Urban Revolution.* London: Zed Books.

Patel, V. (2013) Sanitation and dignity. *The Hindu*, 24 September. www.thehindu.com/news/national/other-states/sanitation-and-dignity/article5160791.ece (accessed 25 August 2016).

Peake, L. (1993) 'Race' and sexuality: Challenging the patriarchal structuring of urban social space. *Environment and Planning D: Society and Space.* 11: 415–432.

Peake, L. (2015) The twenty-first century quest for feminism and the global urban. *International Journal of Urban and Regional Research.* 40(1): 219–227.

Phadke, S., Khan, S. and Ranade, S. (2011) *Why Loiter? Women and Risk on Mumbai's Streets.* New Delhi: Penguin.

Pinto, R. (2016) Right to Pee activists return award as toilets still unclean. *Times of India*, 9 March, http://timesofindia.indiatimes.com/city/mumbai/Right-to-Pee-activists-return-award-as-toilets-still-unclean/articleshow/51318710.cms (last accessed 10 May 2016).

Ranganathan, M. (2015) Storm drains as assemblages: The political ecology of flood risk in post-colonial Bangalore. *Antipode.* 47(5): 1300–1320.

Robins, S. (2014). The 2011 toilet wars in South Africa: Justice and transition between the exceptional and the everyday after apartheid. *Development and Change.* 45(3): 479–501.

Sarkar, A. (2016) Mumbai draft DP proposes adding storeys to public toilet blocks. *Indian Express*, 2 May. http://indianexpress.com/article/cities/mumbai/mumbai-draft-dp-proposes-adding-storeys-to-public-toilet-blocks-2779777/ (accessed 10 May 2016).

Satterthwaite, D. and Mitlin, D. (2014) *Reducing Urban Poverty in the Global South.* London: Routledge.

Schnitzler, A. (2013) Traveling technologies: Infrastructure, ethical regimes, and the materiality of politics in South Africa. *Cultural Anthropology.* 28(4): 670–693.

Scott, J.C. (1998) *Seeing Like a State: How Certain Schemes to Improve the Human Condition Have Failed.* Yale: Yale University Press.

Shah, S.P. (2014) *Street Corner Secrets: Sex, Work, and Migration in the City of Mumbai.* Durham, NC: Duke University Press.

Sharma, R.N., and Bhide, A. (2005) World Bank funded slum sanitation programme in Mumbai: Participatory approach and lessons learnt. *Economic and Political Weekly*, 23 April. pp. 1784–1789.

Silver, J. (2014) 'Incremental infrastructures: material improvisation and social collaboration across post-colonial Accra'. *Urban Geography.* 35(6): 788–804.

Social Justice Coalition (SJC) (2014) *Our Toilets Are Dirty: Report of the Social Audit into the Janitorial Service for Communal Flush Toilets in Khayelitsha, Cape Town.* Cape Town: SJC/Ndifuna Ukwazi.

Thieme, T. (2017) Navigating and negotiating ethnographies of urban hustle in Nairobi slums. *CITY.* 21(2): 219–231.

Truelove, J. (2011) (Re-)conceptualising water inequality in Delhi, India through a feminist political ecology framework. *Geoforum*. 42: 143–152.

UN Millennium Project (2005) *Health, Dignity and Development: What will it take? The Report of the Millennium Taskforce on Water and Sanitation*. London: Earthscan.

WaterAid (2007) *The State of the World's Toilets, 2007*. www.wateraid.org/documents/the_state_of_the_worlds_toilets_2007_1.pdf (accessed May, 2008).

World Health Organization (WHO) (2012) GL*ASS 2012 Report - UN-Water Global Analysis and Assessment of Sanitation and Drinking Water: The Challenge of Extending and Sustaining Services*. Geneva: World Health Organization.

Yardley, J. (2012) In Mumbai, a campaign against restroom injustice. *New York Times*, 14 June. www.nytimes.com/2012/06/15/world/asia/in-mumbai-a-campaign-against-restroom-injustice.html?_r=0 (accessed 16 August 2016).

4 Enframing citizenship

Social housing and ontological orientations in Johannesburg

Alex Wafer

Introduction

On 8 April 2015, just before midday, a spontaneous community protest erupted in Cosmo City, a public housing development north of Johannesburg. The protest made the local news after a policeman was injured in a scuffle, and several shops in the area were damaged by protesters, notably the local KFC (Hartleb 2015). But aside from these incidents, the protest passed with relatively little public attention. Community-based protests are not unusual in South African towns and cities over the past decade, most often directed against the perceived failures of municipal and provincial infrastructure delivery.[1] The occurrence therefore of a spontaneous community protest in a low-income neighbourhood on the outskirts of Johannesburg was not unusual in its generality, but it was surprising in its specificity for two reasons: firstly, the protest took place in a recently completed public housing project, celebrated in the media and by the national and provincial housing departments as a model for post-apartheid community building. Certainly, unlike the largely homogeneously poor, under-serviced and geographically peripheral former black townships, as well as many of the early post-apartheid public housing projects which were built on cheap ex-urban land, Cosmo City is relatively well-located (part of a strategic growth corridor to the city's second airport), it is socio-economically vibrant and well-serviced in terms of schools and community facilities.

The second reason why the protest was surprising is that it was directed *against* municipal intervention; specifically against attempts by the local authorities to enforce building regulations through the removal of unlawfully constructed backyard rooms, i.e. rooms built for rental in the backyards of houses, usually outside of municipal building regulations (Shapurjee and Charlton 2013). Whereas in many communities and neighbourhoods across the country residents mobilise to express feelings of abandonment by the state, in Cosmo City residents felt that the state, having provided them with promised houses, was now over-stepping its involvement in their lives and livelihoods. This is significant in the context of recent research and writing on cities of the global South, which has in general focused on the emergence of demands made on state institutions to be increasingly present in the lives of political subjects (see e.g. Agarwal 2004; Chatterjee 2004; Roy 2009). But the form that this presence takes is the context of conflict: to the extent that the state is here imagined as a homogeneous entity it is not invoked as the guardian of abstract legal rights; rather it is invoked as gatekeeper of substantive material benefit, usually in the form of infrastructure provision (Anand 2011; Chatterjee 2004). The protest in Cosmo City suggests that emergent imaginaries of citizenship can also include the disciplining and/or the eschewal of state institutions, just as these institutions seek to contain emergent expressions of citizenship (see also Chance 2015 for a different articulation of this point).

Events like the protest in Cosmo City can serve as a useful lens through which to understand dynamics beyond the specific context of the event itself. Morton Nielsen (2010) reminds us of Max Gluckman's (1940) analysis of social situations not as apt examples of social or cultural practices, but rather as exposing the contradictions within apparently stable social orders. Kapferer (2010) reiterates the "unstable equilibrium" of social situations, suggesting that the apparent durability of social order resides in perpetual fluidity and instability. While Gluckman was interested in the possibilities of endurance of social order despite internal crisis (McMilllan 1995), Nielsen and Kapferer are both more interested in the potentialities that unstable equilibrium contains. Nielsen (2010) proposes that we read such situations through the Deleuzian concept of "event" as: "neither a decisive rupture nor a new beginning (…) but the introduction of change and variation into already existing structures" (Nielsen 2010: 155).

I suggest that the protest in Cosmo City may be usefully understood from this eventful perspective. While on the surface the protest was about angry residents halting the removal of unlawfully built backyard rooms, the event in fact exposed a disjuncture regarding the ontological status of the very infrastructure of social housing: i.e. the so-called "RDP house" (in reference to the Reconstruction and Development Programme which constituted the initial post-apartheid housing programme). On the one hand, the RDP house is the infrastructure through which subjective relationships with the post-apartheid state are gathered together and re-composed, approaching what Appadurai and Holston (1996) refer to as the substantive realisation of citizenship (rather than merely juridical citizenship).[2] On the other hand, the RDP house represents an imperfect but nevertheless ready-to-hand object onto which expectations of becoming can be projected. As the protest reveals, beneficiaries of RDP houses do not always orient themselves in respect to those infrastructures in ways that the institutions of state expect or even approve of. To better understand the event of the protest, and its deeper ontological disjuncture, requires that we attend to the multiple ontologies of the RDP house. It is to that which the remainder of this chapter is directed.

The ontologies of infrastructures are the preoccupation of a growing body of literature, especially focused on the city in the global South. In a well-cited essay, Larkin (2013) suggests that the ontology of infrastructure resides in the fact that infrastructures "are things and also the relation between things" (328). Larkin suggests at least two ways in which infrastructures become significant: in their techno-political and in their poetic dimension. By techno-political Larkin refers to the bio-political rationalities that are embedded in infrastructures and which constitute them as an "apparatus of governmentality" (Larkin 2013: 328; see also Anand 2011; von Schintzler 2008). By poetics he refers to the semiotic and aesthetic modes through which infrastructures address subjects. One important outcome of this literature is the recognition that the site of politics often lies in these infrastructural networks, and that infrastructural networks are seldom merely technical networks. While this has been a very fruitful direction in the literature, there has also been a tendency to read the ontology of infrastructure from its techno-political or semiotic rationalities, at least implicitly. Moreover, Cesafsky (2017) has cautioned the tendency towards a normativity in which particular infrastructural forms and distributions are assumed to produce a more equitable social order (see e.g. Kooy and Bakker 2008). While this literature has highlighted the implication of infrastructures into social and political processes "it excludes from view a vast array of elements that are crucial to infrastructural flows" (Cesafsky 2017: 152). While Cesafsky might make this point rather too strongly, I would agree that there is an implicit assumption that subjectivities are produced in the assertion of – and resistance to – the rationalities contained within infrastructural form. I would argue that much of this literature does not provide sufficient theoretical tools for understanding the more complex implication of infrastructures into a social and material presence.

A more recent branch of the literature on infrastructural ontologies has explicitly emphasised the uncontainable materialities of infrastructural networks, the instability, failure, and constant reconfiguration that is constitutive of social and material worlds, and the ways in which infrastructural networks are always exceeded by the multiplicity of endurance. Much of this work has emerged on the city in the global South, where the instability of material and social presence is perhaps more readily observable than in many cities of the global North (see e. g. De Boeck 2013; Le Marcis and Inggs 2004; Simone 2008; Wilhelm-Solomon 2017). Much of it (though by no means all) is influenced by the so-called new materialism of assemblage theory, and Actor Network Theory (Cesafsky 2017; McFarlane 2011). McFarlane's (2011) engagement with assemblage theory with regards the material instability of the present is particularly productive. Yet, while this diverse body of literature has highlighted what Gordillo (2014) refers to as the "ruptured multiplicity" (i.e. the multiple alternate imaginaries of political and ontological subjectivity) of the material and social present, it has perhaps under-emphasised the persistence and obduracy of particular infrastructural assemblages. This is the case with the RDP house, the form of which reproduces materialities of inequality that have characterised South African cities from apartheid into the present. The site of Cosmo City is here illustrative, the development itself having been initially planned as a racially segregated township in the dying days of apartheid.

In this chapter I seek to account for the "ruptured multiplicity" of subjective orientation in Cosmo City in relationship to the materiality of its presence. I consider the ways in which particular infrastructural assemblages in a particular context persist over time (their durability), in the process containing and constraining the possibilities of ontological becoming. At the same time, I consider the ways in which these infrastructural assemblages are open to failures, short-circuits, over-flows and re-compositions (their mutability) even as their material form persists. In the following section I discuss the emergence of the RDP house as the dominant form of social housing in South Africa, tracing its emergence to forms of apartheid-era urbanism. I propose specifically that Heidegger's (1977) concept of "enframing" might be usefully revisited here – albeit through a more immanent lens. In the following section I then trace the emergence of Cosmo City as a particular example of a post-apartheid social housing project, again tracing the stubborn persistence of apartheid-era urbanism. In the final section, I look at the ways in which individual beneficiaries have used this ultimately imperfect material presence towards forms of becoming: part interpellated by the rationalities of the prevailing infrastructural assemblage, part in resistance to it, and part exceeding or subverting the terms of its address.

The durability of infrastructural form

The abiding legacy of the now-defunct Reconstruction and Development Programme (RDP) which constituted the original post-apartheid social-housing programme, the eponymous "RDP house" has been reproduced over three million times across the South African landscape through various iterations of housing policy (Huchzermeyer 2014). For some commentators its apparent success as a typology is qualified with respect to the quality of the houses built (apartheid-era township houses are claimed to be more robustly built), their location (on greenfield sites far from viable economic opportunities) and the fraught process of delivery and allocation (early experiments with subsidised self-built initiatives quickly gave way to large-scale developments contracted to corporate property-developers) (Gabula 2012). Partly in recognition of these criticisms the housing programme was repackaged in 2004 as *Breaking New Ground* (BNG), the emphasis of which (in discourse if not so obviously in practice) was a shift away from quantity towards creating better and more sustainable communities (Lemanski 2009). More recent policy discourse has reverted once again to the economies of scale of large-scale provision, although the rationalities have changed somewhat towards the goal of "catalytic" urban developments (Ballard 2017). Shifts in discourse and imperative over the past several decades notwithstanding, the basic element of low-cost state-subsidised social housing in South Africa has remained the direct transfer of title for small free-standing bungalow units to wait-listed beneficiaries. Despite criticisms of their material instability, their geographic peripherality and their rudimentarity, the RDP house has directly impacted the lives and livelihoods of millions of South Africans. It is of course not the only form of state assistance that poor South Africans are eligible for: over 13 million South Africans receive social grants in the form of state pensions, child support grants or disability grants, all of which arguably have greater financial impact on the lives and livelihoods of poor South Africans (Turok and Borel-Saladin 2016). But the RDP house, more than any other state expenditure, manifests in material form the presence of the post-apartheid state in the everyday lives of South Africans. It is therefore a powerful technology of governmentality, extending the reach of the state into the everyday lives and imaginaries of its direct (and indirect) beneficiaries (I will discuss these secondary beneficiaries in the final part of this chapter). Indeed, when residents of Cosmo City were relocated from a nearby informal settlement, they were offered lessons on how to inhabit a formal home and how to live in formal urban communities (Mail and Guardian 2007). Charlton (2018) has however suggested that governmental rationalities manifested in the RDP house contain an imaginary of citizenship that does not always match those enacted by their beneficiaries, who have used the house in ways that it was not originally designed for. She argues that the governmental rationality undergirding the RDP house is premised on a deliberate un-knowing, and thus avoids the need for a critical self-reflection on the actual impact on lives and livelihoods. Bernard Dubbeld (2017), drawing on Chatterjee's (2004) distinction between the government of civil and political society, suggests similarly that the rationality of the RDP house is not necessarily the realisation of full citizenship but rather the management of the expectations of a population through benefits that "never quite become rights" (Chatterjee quoted in Dubbeld 2017: 223). Like Charlton, Dubbeld suggests that beneficiaries contest ascribed forms of subjectivity and actively reclaim citizenship through practices of what Lemanski (2009) calls "augmented informality" (472).

Tim Mitchell (1988), tracing the massive transformation of political subjectivity in Egypt in the early twentieth century, identified similar patterns of techno-political governmentality through the extension of new schools, new urban settlement typologies and new semiotic registers of national identity. Mitchell borrowed Heidegger's (1977) concept of "enframing" to account for what he saw as the wholesale ontological recalibration that this social revolution implied. Translated from the German word *Gestell,* Heidegger's enframing refers to the non-technical and non-technological ontology that resides within what is called technology: "enframing means that way of revealing which holds sway in the essence of modern technology and which is itself nothing technological" (Heidegger 1977: 20). Heidegger's concern was that technology is not simply a more efficient means to an end, but exists as a mode of ordering the world. For Mitchell, this mode of ordering was part of a deliberate strategy through which the Egyptian state intervened directly into the lives of its subjects, although I would argue that Heidegger did not conceptualise enframing as such a capacity to be deployed (even as certain institutions and assemblages may be directed towards the homogenization or "fixing" of ontology). Rather enframing is an a priori condition within which capacities, technologies and imaginaries can be thought in the first place. This a priori condition can nevertheless be revealed and contested: indeed this is for Heidegger the role of philosophy, and an objective of his own ontological phenomenology. Heidegger's specific preoccupation was with the ways in which modern technologies are premised on an ontology of utility, but the concept of enframing as referring to a mode of ordering pertains (at least in principle) to any ontological framing of the world – i.e. enframing is a fundamental feature of the possibility of thinking about the world.

The concept of enframing is useful here, I argue, because it allows us to account for the durability of particular infrastructural assemblages and forms, even though the rationalities that undergird them may have changed. Mitchell's (1988) use of the terms was applied to the total "governmental apparatus" that was directed towards the territory and population of Egypt, first by the British colonial state and later by the post-colonial nationalist state. Mitchell draws heavily on Foucault's theorization of "disciplining mechanisms" (1988: 35) as a way to interpret the concept of enframing – as ontological re-ordering towards the rationalities of discipline. Notwithstanding Mitchell's powerful argument, I suggest that he under-states the ontological disjuncture that the governmentalisation of Egypt implied, and over-states the equation of governmental rationality with ontological re-ordering. My reading of Heidegger's (1977) original concept is that ontology exists as a kind of meta-cognition: framing the conditions for thought rather than a rationality that can be deployed or instantiated. Heidegger himself suggested that we cannot affect enframing's removal, but he did suggest that the task of philosophy should be to prepare ourselves for its change.

At the same time, I am concerned that Heidegger's original conceptualization of the term is itself somewhat limited by its assumption of ontological hegemony. Though Heidegger offers what Turnbull (2009) refers to as a "redemptive of ontology" rather than a simple critique of technology's corrosive potential in the present, he nevertheless does assume a total grip on our ontological relation to reality (Brock 2003). Against this totalizing ontology, Matthew Wilhelm-Solomon (2017) has proposed what he terms a multiplicity of "ontological orientations [which involve] attempts to interpret, stabilise and reconfigure relations of existence through embodied and material practice" (2017: 174). Wilhelm-Solomon draws on a wide range of thinkers, including inter alia Descola (2013), Gordillo (2014) and Povinelli (2016), to argue for a far more multiple and multiplying conceptualization of ontology. But if the unstable materiality of presence ruptures in multiplicity, how might we account for the persistence of particular material form? Gordillo (2014) draws widely on Walter Benjamin and Theodor Adorno in his discussion on rubble, but in particular their constellational ontologies. Gordillo begins from the premise that we live in a ruinous world, in which the instability and entropic decay of our material presence is constitutive of our experience of that presence. But to make sense of this material instability requires that we recognise the consistency of form. As Gordillo quotes Adorno: "cognition of the object in its constellation is cognition of the process stored in the object" (Adorno, quoted in Gordillo 2014: 20). This constellational ontology – not dissimilar to e.g. McFarlane's (2011) engagement with assemblage theory – allows us to recognise the historical and social processes embedded in the durability and obduracy of particular form, even in the context of dissipating multiplicity. I suggest that, when read through this ontological multiplicity, the concept of enframing remains a useful concept in understanding how these obdurate forms continue to shape ontological orientation, without necessarily settling or fixing ontology. The case of the RDP house is here illustrative: although the object of the house decays and mutates, its typological form persists and is endlessly reproduced. And this persistent and obdurate form enframes the possibility of aspiration and expectation.

South Africa has a deep and abiding obsession with the single-story suburban bungalow. In part, this fetishisation reflects rural utopic imaginaries – albeit very differently for different ethnic and racial groups (Chipkin 1993). In separate part, the ubiquity of the bungalow-imaginary reflects the industrialization and expansion of South African cities in an era of global Fordist capitalism, so that some have referred to South African urbanization in the twentieth century as racial Fordism (Rogerson 1991). The bungalow-house with the swimming pool, the lawn and the pink flamingo post-box became in many regards the white South African version of suburban utopia in the 1960s and 1970s, and barely masked a deep antipathy towards urban modernity (Chipkin 1993). Unlike in Brazil, for example, where "tropical modernism" was embraced as the style of the future, in South Africa the skyscraper has long been associated with urban anxiety. In part this antipathy can be traced back to long-held resentment by white Afrikaners towards the predominantly urban-based English-speaking South Africans, whose presence represents the loss of homeland in the face of British Imperialism. While Afrikaner mythology memorialises their protestant and agricultural roots, the exploitation of the gold-fields by British Imperial capital in the late nineteenth and early twentieth century brought the growth of cities like Johannesburg, associated with vice and avarice, and massive populations of "poor whites" – i.e. those Afrikaners forced off agricultural land and into urban slums in the 1920s and 1930s.

In spite of the gradual urbanization of the economy in the twentieth century, the suburbs remained a respite for many from the worst aspects of the city. While white South Africans sought voluntarily dissociation from urban centers into the ever-expanding suburbs, black South Africans living or arriving in the city were forced out of the city. A famous photograph in the Apartheid Museum shows five white men dressed in 1950s attire examining a map entitled "native townships"; they are City Councillors assessing the location of the future Soweto. Findley and Ogbu (2011) and Nicholas Coetzer (2009) separately describe how these plans were influenced by international trends in suburbanization, especially the garden-city concept of Ebanezer Howard. But the racist ideology of apartheid meant that they were realised with a meanness that found its form in the so-called "matchbox houses" that became synonymous with the indignity and inequality of townships. It is in this object of racial distinction, ironically, that the form of the future RDP house resided.

In the post-apartheid era, the typology of the bungalow has been perpetuated at both upper and lower ends of the economic spectrum, as low-rise gated communities and private estates on the one hand, and RDP and low-income housing developments on the other, push the urban edge ever outwards (Mabin et al. 2013). This is driven partly by a political-economy in which peripheral land is cheap, the building industry is monopolised by a few corporate property developers, building quality is low, and housing variations are limited (Butcher 2016). But it is also reproduced through the ways in which this persistent typology of the bungalow frames aspirations and imaginaries of becoming in the present. And as the small trickle of South Africans who have managed to make the slow journey of class mobility seek to instantiate themselves in the urban landscape, or as the institutions of state seek to address those poor who present themselves as subjects awaiting governmentalisation, it is the suburban bungalow rather than the urban skyscraper that is the constellational form through which these aspirations materialise. In the following section I will look at one very particular site in which this constellation converges: Cosmo City.

Cosmo City and the township legacy

In his essay "Building Dwelling Thinking" Heidegger (1971) makes a connection between dwelling and the building of communities: "we attain to dwelling, so it seems, only by means of building. The latter, building, has the former, dwelling, as its goal (…) [but] building is not merely a means and a way toward dwelling – to build is in itself already to dwell" (unpaginated). Though sometimes opaque in his writing (and made often more so in translation), the essay attempts to account for belonging through the material forms that *gather* together and constitute the social, imaginative and ontological constellations through which we endeavour to create durable social lives. The concept of enframing is clearly related here, because this gathering together – this constellation of material, social and ontological – endures and persists. Indeed, this endurance is necessary for social life. In this section I want to consider the *gathering* that is Cosmo City. In doing so I will demonstrate how the enframing ontology of the bungalow house is generative of this particular urban form as the constellation of inter alia changing governmental imperatives, recuperative justice and imaginative futures.

Envisaged since at least the early 1990s as an integrated multi-racial and mixed-income community, Cosmo City took almost twenty years to be realised, in the process morphing into something as yet unformed. Envisaged during apartheid as a segregated township to house black South Africans at the periphery of the white city, and then in the 1990s as the possibility for an experimental non-racial suburban community, the final product reflects uncomfortable compromises between social ideal and urban pragmatics. Its material form bears unflattering comparison to the urban form of apartheid-era townships; the distribution of commercial facilities has reproduced homogeneous landscapes of consumption dominated by large monopolies rather than sustaining local economic opportunities; and the organisation of housing typologies has consolidated class-based mistrust between neighborhoods within Cosmo City. Nevertheless, in unanticipated ways Cosmo City has begun to produce new articulations of citizenship, connected in part to the materiality of the housing typologies from which it is primarily constituted.

Officially opened in 2008 (although the first occupation of RDP houses began in November of 2005) it was the first large-scale mixed-income and mixed-ownership public-private housing project built in the post-apartheid era (Haferburg 2013). The low-density development sprawls across 1200 hectares of former agricultural and wetland, with suburban-scale relationships between homes and public amenities. Despite the fact that it was designed to accommodate mostly poor and lower-middle class inhabitants, there is little that is realistically accessible on foot. Although there is a public transport terminal on the eastern edge of the development where the central boulevard intersects with an adjacent freeway, there is no internal public transport which means that residents must walk several kilometers to access the terminal. This is exacerbated by the fact that RDP houses are located in the core of the development, ostensibly so that they would be most well-located to community resources but actually making these poorest residents the most marginalised from economic opportunities outside of the development.

The project was financed via a public-private partnership between the City of Johannesburg, the Gauteng Provincial Department of Housing, and a consortium of local and international financiers and developers (although the sunken costs of the project include decades of planning and un-planning). The final project was intended to provide mixed-income housing for up to 12,000 households, half of which would be RDP houses. In addition to homes, the development included allowance for 12 schools, a community center, several open public parks and undeveloped sites allocated for future development of leisure spaces, commercial developments and potential future expansion of the development. These undeveloped sites remained the property of the developer, which continues to oversee the management and overall development of the project. As of 2018 Cosmo City is home to over 80,000 people, representing on the one hand a massively successful project of community building, but on the other hand densities of occupation that far exceed what was initially planned for (Haferburg 2013).

Its origins go back to the 1970s, when the land was first identified for the development of low-cost housing by the apartheid government – initially called Noweto in counter-reference to the township of Soweto to the southwest of the city. At that time the land was still commercial farmland belonging to several white farmers, but was increasingly well-located to the expanding white suburban municipality of Randburg, north of Johannesburg. The rationale for a township development north of the city had to do with the changing spatial economy of the city from the 1970s onwards: away from the former mining and heavy industrial South towards the expanding financial and services sectors along the Johannesburg-Pretoria corridor (Beavon 2004; Beall et al. 2014). This rationale began to shift in the 1980s: as the state's ability to sustain the edifice of apartheid began to collapse, people began to move into formerly segregated cities, initially as a form of quiet encroachment (Bayat 2013) but by the late 1980s associated with violent anti-apartheid protest. Large "informal settlements" developed north of Johannesburg, notably the settlements of Zevenfontein and Riverbend, which were characterised by over-population, violence and broader public health risks. By the early 1990s endemic violence had taken root in Zevenfontein and Riverbend which echoed violence in the country more broadly – i.e. between groups supporting a national democratic resistance against apartheid and those agitating for ethnic separatism. To mitigate this rapidly deteriorating urban crisis, residents were promised housing in a future Cosmo City; however, by this point the apartheid economy and its political legitimacy were in terminal decline.

Anticipating the demise of apartheid, the municipalities of Johannesburg and Randburg began to plan for a future non-segregated "free settlement" housing development to accommodate new urban migrants (Johnson 1990). But beyond the idealistic aims of an experimental non-racial community, the primary political motivation for the development of Cosmo City was to relocate residents from Zevenfontein and Riverbend. The first phase of relocations, in 1993, moved people to a hastily built housing development further north of Cosmo City, called Diepsloot. Although not built under the RDP programme, its hastily constructed and rudimentary form would set a precedent for future RDP projects. Diepsloot has been progressively expanded over the following decades, with several new extensions being explicitly delivered under the RDP programme, to become a densely populated area of social housing and informal infiltration (Harber 2011). While infinitely more livable than the original informal settlements of Zevenfontein and Riverbend, nevertheless Diepsloot stands as the exemplar of everything that Cosmo City would not become: homogeneously overcrowded, under-serviced and geographically peripheral.

As part of the provisional peace agreement with residents of Zevenfontein, one group who refused removal to Diepsloot was promised houses in a future Cosmo City. But by the late 1990s the development was still beset by delays. Initially, the farm owners had refused to sell the land. Then, when the land was finally secured, local white residents of neighbourhoods close to the planned Cosmo City development lodged court orders against the municipality and the developer to contest the development, citing the impact on property values, urban congestion and a host of other nimby-ist concerns (Haferburg 2013). In retaliation, it is claimed (based on anecdotal reports from research participants) that the residents of Zevenfontein would periodically appear on the farms and put stakes in the ground demarcating their future houses – an intimidation tactic to encourage the farmers to sell. By the time the first bulldozers were on site in the early 2000s, both the "northern township" and the subsequent "model community" imaginaries had been replaced by new social imperatives. South African housing policy had shifted from the RDP to the "Breaking New Ground" policy, which emphasised sustainable cities. Now the development was seen as a pioneering prototype for sustainable community-building, woven into the wetland ecosystem that existed on parts of the land as well as the emerging economy connected to the city's second major airport hub, and all homes to be fitted with solar water-heaters (Urban Landmark 2011). But the original blueprint, of small bungalows on little suburban plots, remained the basis of imagining the development.

Travelling into Cosmo City in 2017 reveals mixed successes. For the most part, the neighbourhood appears as a conventional suburb: some of the saplings that were originally planted along the streets have grown into substantial shade-bearing trees – though not many. The black-top is mostly smooth and without pot-holes, but the streets are dusty and wide and cars drive too fast. Large parts of the RDP sections are characterised by informal economic activity, but the formal economic nodes are made up of petrol garages and large-scale retail stores selling cheap consumer items; an economic model out of scale with the granular qualities of sustainable communities. While the development was intended to house 50,000 people, the actual number is closer to 80,000 with people living in backyard rooms or renting rooms in their homes (Kriegler 2016). In the final analysis, Cosmo City appears to have reproduced – albeit with decent social services planned in – the old segregated township model. Despite the social resources invested into the development, the "model community" has failed to produce an imaginary outside of the long tradition of segregated suburbanism. But the point here is not to offer a normative reflection on Cosmo City, nor is it to claim some South African exceptionalism in this suburban form. Rather, it is to demonstrate the persistence of form within which imaginaries of becoming can even exist. However, the presence of Cosmo City is not only the unsatisfactory realization of poorly considered governmental rationalities, somehow diluted by the persistent logic of older urban form. The development has been adapted and reworked, so that its outward form may tell us little about the lives that inhabit the space. In the final section I will look at the material mutability of the RDP house, and the possibilities the RDP house offers for alternate imaginaries of becoming.

Mutable citizenship

In the previous two sections I argued that Heidegger's concept of ontological enframing offers a way of thinking about the obduracy of infrastructural assemblages, despite changing political rationalities. I showed firstly how the object of the RDP house reflects and in some way reproduces apartheid-era typologies despite new imaginaries of post-apartheid citizenship projected onto it. I then showed how, just as the RDP house reproduced the township matchbox house writ small, Cosmo City reproduces the township form writ large. In this final section, I consider the ways in which beneficiaries have actually engaged with the RDP house (and with the materiality of Cosmo City) in ways which contradict these ascribed subjectivities, specifically through what are called backyard economies. I focus on the mutable qualities of the RDP house within the enframing logic of the infrastructural assemblage and show how even these alternative ontological orientations are enframed within the imaginary of the suburban bungalow. But I suggest that backyard economies represent the manifestation of alternative imaginaries of being and becoming through the imperfect though ready-at-hand RDP house.

Backyard economies have a long history in South African cities as an ad hoc response to restricted accommodation options and livelihood opportunities among poor urban residents (Huchzermeyer 2014; Lemanski 2009; Lategan and Cilliers 2016; Sharpujee and Charlton 2013). Leslie Bank (2011) has shown how constraints on black mobility in urban spaces under apartheid led to severe overcrowding in township areas, and the over-saturation of the few functional infrastructures available to black urban residents. Informal tenure practices and informal economic activities allowed black South Africans to maintain an urban foothold in the context of economic decline and political instability in the 1980s. This barely durable presence of black South Africans not only represented an urban crisis in itself, but provided the conditions for urban-based protest and the eventual collapse of apartheid. As we discussed above, the initial conceptualisation of Cosmo City was one attempt to respond to this urban and political crisis.

In the post-apartheid context backyard economies – like informal economies more broadly – have expanded in interesting ways, extending also into many other parts of the city including former white-suburban neighbourhoods and newer RDP neighbourhoods. Those engaged in such economies are for the most part not self-reflexive about their activities, but in the face of haphazard municipal sanction residents of Cosmo City defend the practices as a legitimate form of capital accumulation in a context of historical disadvantage and currently high levels of unemployment (Charlton and Kihato 2006; Shapurjee and Charlton 2013; Turok and Borel-Saladin 2016). Backyard accommodation is a significant sector of the housing market among the poor and precarious who are otherwise not accommodated by formal housing and rental markets in the city. Lemanski (2009) reads the technically illegal practice of backyard rentals as a response to a housing policy which has predicated ownership over other social housing options. While the vast majority of backyard accommodation is not technically compliant with municipal building regulations – sometimes consisting of little more than a wooden hut, other times consisting of significant multi-storey constructions – backyard renting represents for tenants a safer and more stable alternative (in terms of both personal safety and tenure) to the under-serviced, legally precarious and crime-vulnerable informal settlements.

During 2017 the author and several student researchers from the University of the Witwatersrand's School of Geography, Archaeology and Environmental Studies undertook several months of field research in Cosmo City. The primary aim was to understand the emergence of new class identities in RDP settlements. In particular, we were interested in relations between two groups: the first are the original RDP beneficiaries who draw on very particular experiential repertoires to reconstitute the RDP house not as a tool of governmentality but as a tool with which to discipline the state. This was evident in the protest with which we began this chapter, but it is a deeply pervasive form of legitimation in Cosmo City. The second group is what I will call secondary beneficiaries, who are those people who rent backyard rooms. Their claim to belonging is much more precarious, but their forms of becoming exceed the material form of Cosmo City; quite literally as their presence is blamed for the burst sewage pipe that sparked the municipal action to demolish backyard rooms, which sparked the initial protest.

The preliminary findings of this research suggest that for RDP beneficiaries the object of the house does not merely connect them into circuits of political subjectivity, but also into circuits of broader civic and financial capital accumulation. While nominally poorer than e.g. owners of mortgage homes, in effect RDP beneficiaries represent an emergent petty landlord class, with forms of social and cultural capital that are of increasing valence in the political economy of increasing informality. In terms of RDP beneficiaries' sense of entitlement regarding the use of their homes there was a strong sense that the RDP home was not a "free" house but one owed them on the basis of either a contractual interpretation of the recent political history, or for some it was seen as a kind of autochthonous right – in terms of the historical legitimacy, the majority (but not all) of the beneficiaries traced some connection to the settlements of Zevenfontein and Riverbend. While the basis of the RDP process envisages a more or less patrimonial relationship with the state – one in which the material form of the house enframes beneficiaries as good and responsible citizens through governmentalising particular urban practices and behaviours – in the case of Cosmo City it is read far more as the state's obligation:

> They say it is a free house but it is not a free house. I had a shack before and they told me to leave that and come to this house. They took the materials from my shack, so this is not a free house – it is my house that was promised me. (Backyard landlord, Cosmo City, interview 14 July 2017)

It is also clear from the interviews that the RDP house is not seen only as a home, but as the less-than-ideal form through which rectification of historical injustice and economic exclusion is framed:

> What do they [the government] want? I have no job, no income, only this little house that they have given to me. They say it is *mahala* [free]. I had a business there [Zandspruit] before, selling old clothes. Now I have a business where I am [a small shop], and I can rent my backyard. (Backyard landlord, Cosmo City, interview 12 July 2017)

This conclusion is borne out by several other studies on RDP beneficiaries and backyard economies (see e.g. Kihato 2014; Lemanski 2017).

Moreover, many RDP beneficiaries with whom we spoke claimed to have lived in either Zevenfontein or Riverbend before they were allocated homes in Cosmo City; others claimed to have parents who had lived in these communities. That claim is not only powerful in its connection to an historical narrative; more specifically most of the beneficiaries we interviewed were members of the local ANC ward committee. The "originary mythology" of Zevenfontein and Riverbend is embedded in promises made to residents during the anti-apartheid struggle. The decades-long duration spent waiting also lends a particular valance to their narrative. Thus, beneficiaries of RDP houses in Cosmo City have a strong sense that they are the legitimate authority in Cosmo City. This originary mythology and the invocation of some form of autochthonous entitlement are clearly embedded in a legitimation of on one hand national political and civil structures, and on the other hand local community-embedded dynamics. This scalar articulation is encapsulated in membership of local ANC branches, which make claims of leadership locally but understood through a national scale/longer duree. Against this scalar articulation is a great deal of animosity towards the meso-scale municipal structures which, although also nominally ANC-led,[3] represent the layer of government of service delivery – and enforcement. This was sometimes articulated as an historical association with apartheid: for most RDP beneficiaries apartheid was experienced at the municipal level, with exclusions from the urban economy and the racial segregation of housing. But it is also entrenched in the structures of delivery in the post-apartheid context. RDP houses are provided by provincial housing departments, on the basis of national housing policy, but services delivery and revenue collection is effected by the local municipality.

Unlike the RDP beneficiaries with whom we spoke, the majority of tenants claimed to be immigrants to the city: from other parts of the country and from outside of South Africa. Most of them do not qualify for an RDP house, or are on a waiting list somewhere else. The average wait time of those who we spoke with was over a decade. As Oldfield and Greyling (2015) note the idea of waiting for the state has become a normalised condition for many South Africans, while backyard rooms have become spaces for what one respondent called "temporary citizenship". As Lemanski (2017) notes, many of these people constitute a missing gap, too poor to participate in formal housing markets yet not poor enough or not eligible to benefit from the RDP scheme. Almost all participants with whom we spoke said that they previously lived in informal settlements; almost half had also at some point lived in or had parents who lived in Riverbend or Zevenfontein. Tenants face several intersecting vulnerabilities. There was a consistent perception that cost of living for tenants was high. This may also have reflected general perceptions of cost of living among what is a poor and vulnerable social group. But there was definitely a sense that Cosmo City represented an expensive option, off-set by the opportunities that the location of the community represents:

> In Soweto I would pay half for a room the same size [...] Cosmo City is very expensive. It is not made for the poor. (Backyard tenant, Cosmo City, interview 12 July 2017)

Or again:

> Cosmo City is expensive, I know in Soweto I would pay less. But Cosmo is also close to work. (Backyard tenant, Cosmo City, interview 10 July 2017)

Speaking much more directly to the issue of access to work and opportunity, another tenant stated:

> I work in houses around Cosmo City and nearby sometimes. I moved here because there is business in Cosmo City. But you have to pay to be part of that business. (Backyard tenant, Cosmo City, interview 14 July 2017)

Despite the perceived cost, tenants moved to Cosmo City for safety and access to better social services: since RDP houses are serviced with water, sewage and electricity most backyard rooms also have access, although there are frequent tensions around access and payment between landlords and tenants. Landlord–tenant relations differed from case to case:

> We have a good relationship with the people in the big house because they know what it is like not to have a proper house in your name. They treat us like family. (Backyard tenant, Cosmo City, interview 12 July 2017)

On the other hand:

> the owner of the house does not respect me because she thinks we are desperate to stay here. She treats us like her children with a time to come home. (Backyard tenant, Cosmo City, interview 14 July 2017)

Yet while these ambivalent and various experiences might characterise the varigated reality of most landlord–tenant relationships in any part of the city, there was nevertheless a sense that both landlord and tenant belonged to Cosmo City. Perhaps this represents a pragmatic response to the fact that landlords make money off the poor and the vulnerable: certainly many RDP beneficiaries did not express hostility to immigrants. One even suggested:

> they [foreign nationals] don't have family and friends that come in and out of the yard all day so they also don't come out at night for no reason. They also pay on time with no excuses because they will have nowhere to go if I chase them out. (Backyard landlord, Cosmo City, interview 10 July 2017)

While this statement represents on one hand a worrying potential for abuse of power, it also suggests an urbain pragmatism that might begin to justify the Cosmo in Cosmo City: a pragmatic cosmopolitanism based on emerging articulations of citizenship and belonging that exceed the enframed citizenship of the ideal RDP beneficiary.

The point that I have tried to articulate in this final section is that the mutability of the RDP house – the ability to change its form towards new economies and realities – has been productive of new forms of associational life and forms of becoming that push against the governmental rationalities of the RDP house. This happens nevertheless within the obdurate form of the suburban bungalow house whose durability is premised in many ways on its mutability.

Conclusion: unstable infrastructures of citizenship

I have attempted to show in this chapter that the RDP house in Cosmo City, as a material object, has implicated its beneficiaries (both primary and secondary) into circuits of capital, power and subjectivity in ways which exceed formal articulations of post-apartheid citizenship. These circuits, as so often the case in the contemporary African city, are never stable and consistent but constituted by interruption, over-saturation and informalisation. The governmental rationalities that underlie these infrastructural lives are frequently exceeded, subverted or simply elided. I will conclude the chapter by returning to the protest, from which we digressed in the introduction. The outcome of the protest, beyond the violence that erupted, was that the municipality retreated. It has threatened to return but as of the time of writing no backyard rooms have been demolished in Cosmo City. What has happened is the increase of tension between RDP beneficiaries and backyard tenants on the one hand and the owners of commercial mortgage homes on the other. The latter blame the former for the periodic overflow of sewage onto the main arterial road through the community, South Africa Boulevard. But perhaps more importantly, in the process residents have claimed Cosmo City as the (albeit imperfect) tool for constituting new practices of urban citizenship. Both landlords and tenants were involved in the protests, some to defend livelihoods, some to defend homes. But both to defend the unstable and rudimentary infrastructure of citizenship: the humble RDP house.

Notes

1 In 2016, the website *Africa-Check* contested a 2014 tabloid claim (City Press 2014) which had suggested there were 30 service delivery protests a day across the country in the ten years between 2004 and 2014, concluding from its own media analysis that community and industrial protest action was close to one or two a week on average. More recently (11 July 2018) media-monitoring website *Municipal-IQ* published a briefing suggesting that 164 community-based protests occurred in local municipalities across the country in 2015 – down from 191 in 2014 (MunicipalIQ 2018). Notwithstanding the difficulties of measuring such data, community-based protests are clearly a characteristic of everyday life in South African cities. They manifest localised disaffection with the pace and scale of infrastructure delivery within many poor communities (Lodge and Mottiar 2016), but they also have a much longer history in South African urban politics such that the invocation of anti-apartheid protest memories can underscore claims to place in a post-apartheid context, particularly in traditional working-class communities (Ballard et al. 2006; Robins and von Lieres 2004; Wafer 2008).

2 The South African housing programme (in its various iterations over the past twenty years) transfers title to a so-called RDP house to waitlisted beneficiaries on the basis of financial need. Beneficiaries must register either with the local Municipal Housing Demand Database or the provincial Housing Needs Register. Eligibility criteria include being the following: a South African citizen with ID; over 21 and mentally competent to sign a contract; married or living with a partner, or single and have dependants (single military veterans or aged people without dependents also qualify); earn less than R3,500 per month per household; a first time government subsidy recipient; a first time home owner (www.groundup.org.za/article/everything-you-need-know-about-government-housing/).

3 In fact, since the local elections of 2016 the municipal legislation is dominated by the opposition Democratic Alliance.

Acknowledgements

Thanks are due to Ms. Desire Muzhambi and Ms. Lebo Mkhabela who undertook much of the field work that informed this chapter, as part of their postgraduate studies at the University of the Witwatersrand in 2017. This research was undertaken as part of the "Infrastructures of state and citizenship in post-apartheid South Africa" project (2015–2017), which was funded by the National Research Foundation of South Africa (CSUR grant #93630). An initial research project was undertaken as part of "Revolution room" (2014), a participatory art project of the Visual Arts Network of South Africa (VANSA).

References

Africa Check, 2016. Are there 30 service delivery protests a day in South Africa? 8 June, accessed 23 December 2018, africacheck.org/reports/are-there-30-service-delivery-protests-a-day-in-south-africa-2/

Agarwal, A., 2004. Environmentality: Community, intimate government, and the making of environmental subjects. *Current Anthropology*, 46(2): 161–190.

Anand, N., 2011. Pressure: The politechnics of water supply in Mumbai. *Cultural Anthropology*, 26(4): 542–564.

Appadurai, A. and Holston, J., 1996. Cities and citizenship. *Public Culture*, 8(2): 187–204.

Ballard, R., 2017. Prefix as policy: Megaprojects as South Africa's big idea for human settlements. *Transformation: Critical Perspectives on Southern Africa*, 95(1): i–xviii.

Ballard, R., Habib, A. and Valodia, I., 2006. *Voices of Protest: Social movements in post-apartheid South Africa*. Durban: University of KwaZulu-Natal Press.

Bank, L.J., 2011. *Home Spaces, Street Styles: Contesting power and identity in a South African city*. London: Pluto Press.

Bayat, A., 2013. The quiet encroachment of the ordinary. *Chimumrenga*, 15 February, accessed 23 December 2018, https://chimurengachronic.co.za/quiet-encroachment-of-the-ordinary-2/

Beall, J., Crankshaw, O. and Parnell, S., 2014. *Uniting a Divided City: Governance and social exclusion in Johannesburg*. London: Routledge.

Beavon, K.S.O., 2004. *Johannesburg: The making and shaping of the city (Vol. 9)*. Unisa Press.

Brock, B.R., 2003. *A Theological Examination of Contemporary Deliberation about the Development of New Technologies, with Reference to M. Heidegger, M. Foucault, and G. Grant: Discovering our dwelling* (Doctoral dissertation). London: University of London.

Butcher, S., 2016. *Infrastructures of Property and Debt: Making affordable housing, race and place in Johannesburg* (PhD thesis). Minneapolis, MN: University of Minnesota.

Cesafsky, L., 2017. How to mend a fragmented city: A critique of "infrastructural solidarity". *International Journal of Urban and Regional Research*, 41(1): 145–161.

Chance, K.R., 2015. "Where there is fire, there is politics": Ungovernability and material life in urban South Africa. *Cultural Anthropology*, 30(3): 394–423.

Charlton, S., 2018. Confounded but complacent: Accounting for how the state sees responses to its housing intervention in Johannesburg. *The Journal of Development Studies*, 54(12): 2168–2185.

Charlton, S. and Kihato, C., 2006. Reaching the poor? An analysis of the influences on the evolution of South Africa's housing programme. In U. Pillay, R. Tomlinson and J. du Toit (Eds.), *Democracy and Delivery: Urban policy in South Africa* (pp. 252–282). Cape Town: HSRC Press..

Chatterjee, P., 2004. *The Politics of the Governed: Reflections on popular politics in most of the world*. New York: Columbia University Press.

Chipkin, C., 1993. *Johannesburg Style*. Cape Town: David Phillip.

City Press, 2014. Zuma sidesteps flames, 9 February, accessed 23 December 2018, www.news24.com/Archives/City-Press/Zuma-sidesteps-flames-20150429

Coetzer, N., 2009. Langa Township in the 1920s: An (extra)ordinary Garden Suburb. *South African Journal of Art History*, 24(1): 1–19.

De Boeck, F., 2013. Of rhythm and amalgamation: The knot as form of the urban. *Salon*, 6: 14–17.

Descola, P., 2013. Presence, attachment, origin: Ontologies of "incarnates". In M. Lambek and J. Boddy (Eds.) *A Companion to the Anthropology of Religion* (pp. 35–49). Oxford: Wiley-Blackwell.

Dubbeld, B., 2017. Democracy as technopolitical future: Delivery and discontent in a government settlement in the South African countryside. *Anthropology Southern Africa*, 40(2): 73–84.

Findley, L. and Ogbu, L. 2011. South Africa: From township to town. *Places Journal*, November 2011. https://placesjournal.org/article/south-africa-from-township-to-town/?cn-reloaded=1 (accessed 22 Aug 2018).

Gabula, Z.H., 2012. *Factors Influencing the Construction Project Success Rates of Reconstruction Development Programme (RDP) Housing Projects in the Eastern Cape: A quality perspective: a census study* (Doctoral dissertation; Durban: University of Technology, Durban).

Gluckman, M., 1940. Analysis of a social situation in modern Zululand. *Bantu Studies*, 14(1): 1–30.

Gordillo, G.R., 2014. *Rubble: The afterlife of destruction*. Durham, NC: Duke University Press.

Haferburg, C., 2013. Townships of to-morrow? Cosmo City and inclusive visions for post-apartheid urban futures. *Habitat International*, 39: 261–268.

Harber, A., 2011. *Diepsloot*. Johannesburg: Jonathan Ball.

Hartleb, B. 2015. Cosmo City blocked off after protest. www.news24.com/SouthAfrica/News/Cosmo-City-blocked-off-after-protest-20150408 (accessed December 2017).

Heidegger, M., 1971. Building dwelling thinking. In M. Heidegger (Ed.), *Poetry, Language, Thought* (pp. 141–160). New York: Harper and Row.

Heidegger, M., 1977. *The Question Concerning Technology, and Other Essays*. New York: Harper and Row.

Huchzermeyer, M., 2014. Changing housing policy in South Africa. In J. Bredenoord, P. van Lindert and P. Smets (Eds.), *Affordable Housing in the Urban Global South: Seeking sustainable solutions* (pp. 336–348). London: Routledge.

Johnson, S., 1990. A pure white hole in the black area, *Weekly Mail*, 1 June, accessed 23 December 2018, mg.co.za/article/1990-06-01-00-a-pure-white-hole-in-the-black-area

Kapferer, B., 2010. In the event: Toward an anthropology of generic moments. *Social Analysis*, 54(3): 1–27.

Kihato, C.W., 2014. Lost dreams? Tales of the South African city twenty years after apartheid. *African Identities, 12*(3–4): 357–370.

Kooy, M. and Bakker, K., 2008. Technologies of government: Constituting subjectivities, spaces, and infrastructures in colonial and contemporary Jakarta. *International Journal of Urban and Regional Research*, 32(2): 375–391.

Kriegler, A. (2016). Comfortably cosmopolitan? How patterns of "social cohesion" vary with crime and fear. *South African Crime Quarterly*, 55(1): 61–71.

Larkin, B., 2013. The politics and poetics of infrastructure. *Annual Review of Anthropology*, 42: 327–343.

Lategan, L.G. and Cilliers, J.E., 2016. Towards more compact South African settlements through informal housing: The case of backyard densification in Bridgton and Bongolethu, Oudtshoorn. *Town and Regional Planning*, 68: 12–26.

Lemanski, C., 2009. Augmented informality: South Africa's backyard dwellings as a by-product of formal housing policies. *Habitat International*, 33(4): 472–484.

Lemanski, C., 2017. Citizens in the middle class: The interstitial policy spaces of South Africa's housing gap. *Geoforum*, 79: 101–110.

Le Marcis, F. and Inggs, J., 2004. The suffering body of the city. *Public Culture*, 16(3): 453–477.

Lodge, T. and Mottiar, S., 2016. Protest in South Africa: Motives and meanings. *Democratization*, 23(5): 819–837.

Mabin, A., Butcher, S. and Bloch, R., 2013. Peripheries, suburbanisms and change in sub-Saharan African cities. *Social Dynamics*, 39(2): 167–190.

Macmillan, H., 1995. Return to the Malungwana – Max Gluckman, the Zulu Nation and the Common Society. *African Affairs*, 94(374): 39–65.

Mail and Guardian, 2007. Jo'burg's Cosmo City: A new hope, 14 May, accessed 23 December 2018, mg.co.za/article/2007-05-14-joburgs-cosmo-city-a-new-hope

McFarlane, C., 2011. Assemblage and critical urbanism. *City.* 15(2): 204–224

Mitchell, T., 1988. *Colonising Egypt: With a new preface.* Los Angeles, CA: University of California Press.

MunicipalIQ, 2018. 2018 service delivery protests all-time quarterly record – for immediate release, 11 July, accessed 23 December 2018, www.municipaliq.co.za/publications/press/201807110947026629.doc

Nielsen, M., 2010. Mimesis of the state: From natural disaster to urban citizenship on the outskirts of Maputo, Mozambique. *Social Analysis*, 54(3): 153–173.

Oldfield, S. and Greyling, S., 2015. Waiting for the state: A politics of housing in South Africa. *Environment and Planning A*, 47(5): 1100–1112.

Povinelli, E. 2016. *Geontologies: A requium to late liberalism.* Durham, NC: Duke University Press.

Robins, S. and Von Lieres, B., 2004. Remaking citizenship, unmaking marginalization: the treatment action campaign in post-apartheid South Africa. *Canadian Journal of African Studies/La Revue canadienne des études africaines*, 38(3): 575–586.

Rogerson, C.M., 1991. Beyond racial Fordism: Restructuring industry in the "New" South Africa. *Tijdschrift voor economische en sociale geografie*, 82(5): 355–366.

Roy, A., 2009. Why India cannot plan its cities: Informality, insurgence and the idiom of urbanization. *Planning Theory*, 8(1): 76–87.

Shapurjee, Y. and Charlton, S., 2013. Transforming South Africa's low-income housing projects through backyard dwellings: Intersections with households and the state in Alexandra, Johannesburg. *Journal of Housing and the Built Environment*, 28(4): 653–666.

Simone, A. 2008. People as infrastructure: Intersecting Fragments in Johannesburg. In S. Nuttall and A. Mbembe (Eds.) *Johannesburg: The elusive metropolis* (pp. 68–90). Durham, NC: Duke University Press.

Turnbull, N., 2009. Heidegger and Jünger on the "significance of the century": Technology as a theme in conservative thought. *Writing Technologies*, 2(2): 9–34.

Turok, I. and Borel-Saladin, J., 2016. Backyard shacks, informality and the urban housing crisis in South Africa: Stopgap or prototype solution? *Housing Studies*, 31(4): 384–409.

Urban Landmark, 2011. Urban Landmark land release assessment tool: Cosmo City case study report, accessed 23 December 2018, www.urbanlandmark.org/downloads/lram_cosmo_cs_2011.pdf

Von Schnitzler, A., 2008. Citizenship prepaid: Water, calculability, and techno-politics in South Africa. *Journal of Southern African Studies*, 34(4): 899–917.

Wafer, A., 2008. Scale and identity in post-apartheid Soweto. *Transformation: Critical perspectives on southern Africa*, 66(1): 98–115.

Wilhelm-Solomon, M., 2017. The ruinous vitalism of the urban form: Ontological orientations in inner-city Johannesburg. *Critical African Studies*, 9(2): 174–191.

5 Traveling technologies

Infrastructure, ethical regimes and the materiality of politics in South Africa

Antina von Schnitzler

In July 2011, in the midst of a particularly cold South African winter, a violent protest occurred in Chiawelo, a poorer area of Soweto.[1] For hours, hundreds of residents blocked one of Soweto's main thoroughfares and protested outside local government offices, in the process igniting a car and burning down the houses of two local councillors. Like the many so-called service delivery protests that have been making regular headline news in South Africa in recent years, the protest in Chiawelo bore an uncanny resemblance to the scenes of spectacular violence in the townships during the 1980s. In an apparent effort to dispel such uncomfortable associations, government and African National Congress (ANC) officials were quick to condemn the protests as "acts of anarchy" by "mobs" and "rogue elements". Meanwhile, the media, scrambling to understand what was going on, had identified the cause of the protest as a set of prepaid meters that had recently been installed by the electricity parastatal Eskom. The meters had automatically cut residents off from electricity service, leaving them cold and in the dark.

This was not the first time that prepaid meters emerged as protagonists in large-scale protests. Over the past 15 years, such protests have occurred at regular intervals. During my fieldwork in Soweto and Johannesburg, I had often seen the meters – pulled out from walls and backyards and dropped at the gates of local government offices – take centre stage at demonstrations. A prepaid meter is a small technical device which, apart from measuring networked services such as electricity or water, automatically disconnects users in cases of non-payment. In order to access services, users have to purchase and load up credit tokens in advance either by entering a numerical code or by using a magnetic key or card. In the past two decades, and in a context of neoliberal reforms prescribing "cost recovery" on the one hand, and widespread non-payment of service charges on the other, prepaid meters have become ubiquitous in South Africa. Simultaneously, South Africa has emerged as an industry leader in the development and innovation of prepayment technology and is also a primary exporter of the meters.[2] While prepaid meters have increasingly also been installed in middle class homes, where they are often preferred to untrustworthy municipal bills, for now, the meters are primarily deployed in poorer, historically black townships and informal settlements. Beyond the now standard prepaid cell phones, most residents in places like Soweto are now connected to electricity and water via prepayment devices. "Living prepaid" with an always precarious connection to flows of water or electricity has thus become an increasingly normalised condition.

And yet, at the same time, many residents in Soweto and other urban areas have bypassed their meters, rendering the meters useless and giving residents de facto free water or electricity. As a result, there is an ongoing low-intensity battle between residents tinkering with the technology and utility officials trying to secure it. This article tracks this technical micro-politics involving residents, engineers and utility officials in a seemingly perennial struggle over the enforcement and evasion of payment. I argue that this politics does not take conventional political forms of public demonstration, disagreement or deliberation, but takes shape at the level and in the language of technology itself. Here, questions that were central to the liberation struggle – about the limits, entitlements and obligations of citizenship – are transduced to novel forms, media and idioms.[3] In this context, technologies and infrastructures are not merely *symbols* or *tools for* political expression; rather, technology itself becomes a *political terrain* for the negotiation of moral-political questions that were at the heart of the anti-apartheid struggle and that continue to animate the forms of life left in apartheid's wake.

Building upon recent work in science studies, political theory and anthropology, this article takes the contemporary conflicts surrounding prepayment as a starting point for a broader reflection on the relationship between ethics, politics and technics. Via a genealogical exploration of the travels of the meter on the one hand, and an ethnographic account of the contemporary metering industry in South Africa on the other, I chart the political life of the meter as it is deployed within a diversity of ethical regimes and techno-political assemblages.

A traveling technopolitics

In following the travels of a small technical device and the ethico-political worlds it is shaped by and that it in turn helps shape, I contribute to an emergent anthropological engagement with infrastructure. Rather than neutral means to more substantive ends, this scholarship has foregrounded infrastructures as central to the multiple constitutions of modernity – fashioning socialities, subjectivities, and affective capacities (Larkin 2008; see also Anand 2012; Chalfin 2010; Collier 2011; Elyachar 2012; Kockelman 2010; Mains 2012; von Schnitzler 2008). Specifically, my concern in what follows will be to trace how technical devices are assembled and re-assembled in relation to particular ethical regimes and political projects.[4] I examine how in their very design, such technologies are scripted with and come to reflect specific ethico-political projects, targets and expectations (Redfield 2012). If, as Brian Larkin (2008) suggests, technologies are always "unstable objects", my focus in the following will be on understanding the semiotic-material *work* through which technologies come to inhabit new contexts. I combine what Susan Leigh Star (1999) termed the "ethnography of infrastructure" with a genealogical investigation of the "travels" of technologies across time and space. Such an approach, I suggest, opens up conceptual and methodological space for an exploration of *forms* of the political outside its conventional locations and mediations, whilst focusing our attention on the specific ways in which the technical is always already mobilised within specific ethico-political conjunctures.

While social studies of technology have opened up conceptual space for a rethinking of technology in less deterministic terms, particularly in the focus on the vernacular uses of technology, such approaches have often been less interested in extending these insights to theorisations of the political.[5] Conversely, political theory has often paid little attention to the material, embodied or affective grounds of political action; indeed, the political has often been defined by its location in a public realm unconstrained by the "urgency of the life process" (Arendt 1990: 60).[6] By contrast, recent scholarship has urged a rethinking of liberal-secular accounts of the political, exploring more affective, embodied forms of political subjectivity and focusing on new locations of ethical and political formations that are often below the threshold of visibility of normative conceptions of political action (Connolly 2002; Hirschkind 2001; Mahmood 2005). Similarly, scholars have pointed to the importance of materiality in the constitution of political actors and of political engagement (Barry 2001; Bennett 2009; Mitchell 2011; Braun and Whatmore 2010).

As Partha Chatterjee (2004; 2011) has suggested, such a rethinking of normative liberal accounts of the political is particularly important in postcolonial contexts, in which formal political channels shaped by the colonial legacy are inaccessible to large sections of the population, such as slum dwellers, the informally employed, or the indigenous. Here, political questions are often contested via ostensibly apolitical *administrative* connections to the state, such as the provision of land, housing or basic services. In such contexts, Chatterjee contends, the political is not located in what is conventionally thought of as the political sphere; rather, it is at the register of administration and population where "rules may be bent and stretched" (Chatterjee 2004: 60) that political questions are often de facto negotiated and resolved.

While Chatterjee's theorisation of "political society" enables a rethinking of the *locations* of the political, in the following, I focus on the techno-political *forms* of political society and the multiplicity of its *terrains*. I explore a less visible, material micro-politics concerned with the shaping of subjectivities, ethical dispositions and political agencies, what one might call a politics of non-publics taking shape at the register and in the language of technology. I investigate how infrastructure comes to mediate a diversity of competing ethical projects, political disagreements and subterranean conflicts that often concern central political questions of civic virtue, basic needs, and the rights and obligations of citizenship. In the politics of meters, pipes or wires, infrastructure itself becomes *a political terrain* on which such questions are negotiated and contested. Here, the political circulates and becomes manifest in different material *forms* (Gaonkar and Povinelli 2003).

In the following, I examine these forms via two linked inquiries. First, I trace a genealogy of the prepaid meter as a *techno-political device.* My aim is not to provide an exhaustive social history; rather, I focus on two historical conjunctures at which prepaid meters suddenly proliferated on a large scale and became enrolled within distinct ethico-political projects. First, late 19th century Britain, when the meter was invented to provide gas to the working classes and became integral to the Victorian project of moral reform. Secondly, I track the conceptual and practical labor of translation as the meter moved to late apartheid South Africa, when it came to be deployed as a device of counter-insurgency to end the anti-apartheid rent boycotts. In each moment, I argue, the meters became integral to the constitution of specific techno-political terrains.

In the second part, I ethnographically explore how this techno-political terrain is fashioned in contemporary South Africa. Drawing on my ethnography with engineers in the metering industry, I examine how prepaid meters come to be *re-scripted* as they are deployed in the aftermath of apartheid and in a context of neoliberal reforms.[7] In following engineers' seemingly perennial quest to develop ever more secure metering devices, I describe a technical war of position in which minute technical innovations become crucial tactical moves dependent on engineers' capacity to mobilise local knowledge and interpretive skill.

Tracking the life of the prepaid meter

Much of my fieldwork in Soweto was carried out in Phiri, a poorer section of Soweto next to Chiawelo, where I began this article. In 2003, Phiri had been chosen as the pilot site for Operation Gcin'amanzi ("Save Water"), a controversial multi-year infrastructure project that would convert all Soweto households from unmetered water connections to prepaid water meters. Thus, residents would no longer be charged a monthly flat fee, but a metered rate that needed to be paid in advance to avoid automatic self-disconnection. Over the years, the project has been at the centre of much protest and media controversy, and it also became the subject of a high-profile, if ultimately unsuccessful, constitutional court case in which five Phiri residents challenged the legality of prepaid water meters.[8] Partly because of this multi-faceted opposition, the project is yet to be completed, though most Soweto households now access water via prepaid meters.

While less legally and morally fraught, prepaid electricity meters have been a concern since the 2000s when social movements like the Soweto Electricity Crisis Committee (SECC) protested their increasingly widespread deployment. During Operation Khanyisa ("Light Up"), SECC-affiliated "struggle electricians" re-connected residents to electricity supply if they had been manually disconnected and also bypassed or removed prepaid meters.[9] Thus, they turned illicit reconnections and meter bypasses into a public spectacle and mobilisation tool. In recent years, organised social movement action has declined and, most of the time, prepaid meters are an unremarkable part of life in Soweto. And yet, the bypassing or "bridging" of meters continues silently on a large scale often without any link to social movement organising or a public, "political" language. Indeed, while the protests against Operation Gcin'amanzi eventually subsided and the utility argued that its surveys showed a 98% satisfaction rate, in 2011, utility officials told me that 30–40% of meters had been bypassed and needed to be retro-fitted with new security features.

To have their meters bypassed, residents may individually pay a skilled person – often a utility employee earning a side income – a few hundred Rands, or a skilled family member or SECC-affiliated technician might do it for free. The decision to bypass is determined by multiple calculations – often both pragmatic and moral – and is also often subject to disagreement within households. While the social movements linked the meters to larger concerns about neoliberal reforms, residents often have a multiplicity of concerns that are often less clear cut and more ambivalent. On the one hand, prepaid meters often came with the promise to erase debts (which most Soweto households had accumulated over years of not paying for services) and to prevent the accumulation of new debts in the future. At the same time, the meters often raised larger worries about affordability especially in a context in which, for years, the vast majority of Soweto residents had not paid for basic services and thus money for water or electricity was often not part of household budgets. Electricity is particularly difficult to afford in poor areas of Soweto, where unemployment is high and many households rely on grants and pensions as primary income.[10] A particularly cold winter, for example, can dramatically increase costs for topping up credit and often multiplies cut-offs and blackouts. Importantly, the meter affords no room for negotiation should money to buy credits be in low supply. An unpaid monthly paper bill will not result in an immediate disconnection, and thus enables residents to delay payment until money has come in. The prepaid meter, on the other hand, automatically cuts the connection without regard for context or special circumstances.

But there are also less immediately apparent reasons for opposition to the meters. In South Africa, and in Soweto in particular, infrastructure has never been merely a neutral conduit for the provision of services, but has always been bound up with questions of belonging and citizenship (von Schnitzler 2008). Part of the promise of liberation was an end to the everyday violence of apartheid. Similar to other post-independence projects, the extension and improvement of basic services, from education or health to electricity and water, was seen as central to national reconstruction. Bound up in objections to prepaid meters are thus often larger concerns about the limits of this promise in the paradoxical context of post-apartheid neoliberal reforms, where services are often extended and restricted at the same time (cf. Mains 2012; Ferguson 2007). Prepayment technology in South Africa is thus coded in very specific terms that may differ widely from other places where the technology has been deployed more recently. Indeed, my point here is precisely that technologies are enrolled within ethical and political assemblages in historically specific ways that may or may not "travel" elsewhere and that may shift over time.

Towards the end of my first bout of research in Phiri, I attended South Africa's fourth annual Prepayment Week, a large international industry conference and the only one of its kind. Held as part of the African Utilities Week in Cape Town's imposing International Convention Centre, it brought together nearly eight hundred delegates from five continents to exchange industry news, present innovations, and sell their products. The conference required a shift in perspective. In the exhibition hall, meters were no longer objects of anger and complaint, but rather displayed to promote the desire of their would-be buyers. Removed from dusty township grounds and unconnected to pipes or wires, the meters were neatly lined up according to functionality and level of sophistication – shiny, disentangled objects, next to and in competition with their peers. Here, the social life of prepaid meters was clearly in its commodity phase (Appadurai 1986). But what kind of a commodity was being marketed here? What made the meters valuable in the eyes of manufacturers and prospective buyers?

I had come to the conference in part to learn more about the history of the meters, and to fill the gaps of my archival research. Many of the original developers, it turned out, were still in the business, though today mostly working for multi-national companies. Many had been engineers during the apartheid era, sometimes with previous employment in military and security industries. During conversations with various delegates, I was repeatedly pointed to Peter Clark, who was described to me as the South African "pioneer of prepaid metering".[11] A man in his early sixties and an electrical engineer by training, he had developed the first meters in the mid-1980s and continued to be a central figure in the industry. As we sat on a bench outside overlooking the bustling convention centre, Clark seemed happy to talk about the early trials – South African prepayment technology was an international success story in which he had played no small part.

Like many of the engineers I had talked to earlier in the day, Clark explained to me that the widespread anti-apartheid rent boycotts in the townships during the 1980s had increasingly become a fiscal problem and that engineers had begun to look for "technical solutions" to the crisis of non-payment. Having first seen prepaid meters during travels in Britain, Clark had a meter shipped to South Africa. He described the meter as "a huge metal box" with "very rudimentary technology". It was coin-operated and thus easy to bypass. This, he maintained, had worked in the UK, since, as he put it, "people there are polite". In South Africa, where, he argued, "people get involved in energy theft, and things like this", this meter would not work and would have to be re-engineered from scratch.

Peter Clark's offhand comment about the character of South African and British users sheds some light on the labor of translation required for a technology to be made operational in new contexts, and I will return to this question in the next section. More importantly, his association of prepayment technology and "politeness" encodes a key problematic that goes to the heart of the *kind* of techno-politics I will be concerned with in the rest of this article. While the concept of techno-politics has been productively used in relation to large engineering projects (Hecht 2009; Mitchell 2002), in what follows, I track a techno-politics that operates on a more micro-political terrain and is centrally preoccupied with the relationship between subjectivity, ethical dispositions and the technical.

The "agency of the penny-in-the-slot meter": tracking the technics of 19th century liberalism (London, 1888)

The prepaid meter began its career in Britain in 1888, at a time when the squalor of working-class life had become a matter of rising concern often expressed in the register of public health. During much of the century, working-class homes had remained excluded from networked grids and instead relied on candles, oil lamps and coal for lighting and heating. The darkness of the working-class home was increasingly seen as a problem, in particular given the growing emphasis on public hygiene and concerns about the time spent by workers in better lit and warmer pubs (Daunton 1983). Infrastructures, and the circulation of water, electricity and gas they enabled, were seen as central in the effort to produce the domestic conditions on which a new moral order could be founded (Joyce 2003).

It is in this context of the rise of biopolitics and its technics that the "penny-in-the-slot meter" was invented and dramatically changed the politics of infrastructure. Within ten years, over 60 per cent of Londoners had domestic gas connections via the slot meter and the number of gas consumers increased from two to eight million. As one engineering report noted at the time, many dark working-class areas had thus "become illuminated through the agency of the penny-in-the-slot meter".[12] While the slot meter enabled the integration of the whole city within a networked grid, it simultaneously divided its population into two sets. A first segment, mostly wealthier households, that could be trusted to pay monthly or quarterly bills and hence would have a contractual relationship to the utilities. And a second segment, mostly the working classes, who would have their connection to infrastructure regulated by a technical device, and were thus more precariously located outside of such contractual relations. Unactivated, the meter would automatically disconnect them from the city's flows.

As with other practices of metrology and automation during the 19th century, prepayment technology became closely bound up with distinctly modern moral anxieties about the blurring of human and machinic agency. On the one hand, the automated mechanism of the meters aroused suspicions about the morality of machinic agency. Numerous accounts of the time depict "automatic machines" as "immoral", tempting users to cheat and in turn prone to cheating the users, laying bare the uneasy coincidence of an unprecedented exaltation of the autonomous human agent and her increasing, but surreptitious dependence on technical instruments.

On the other hand, much like the devices of scientific objectivity studied by Daston and Galison (2007), penny-in-the-slot meters were invested with human virtues such as patience, preciseness and continuous alertness; virtues, which ironically most humans were now seen as lacking.[13] The meters – alongside other domestic measuring devices such as thermometers, scales or clocks – were seen as capable of eliciting and shaping certain habits, subjectivities and dispositions. Indeed, they became integrated within a set of larger moral-political concerns that defined the late Victorian era. First, the slot meter, in its ability to extend gas services, was seen as aiding in the "moral upliftment" of the working classes by improving domestic sanitary practices and encouraging domesticity by simplifying the cooking of warm meals and thus discouraging men from frequenting pubs (Daunton 1983). A second concern was the constitution of the worker not only as a producer, but also as a judicious consumer, in particular in a context of the increasing centrality of credit and consumption to economic life. In many reports of the time, prepaid meters were described as devices that would aid in the production of a more rational attitude to spending and accustom the working classes to inhabit a contractual exchange relation.

While the benefits of the meters were widely touted in engineering journals of the time, the actual process of their introduction was beset by problems. Despite engineers' insistence that the meters were accurate, this could never quite be established in the users' mind – to "lie like a gas meter" soon became a popular metaphor to express habitual mendacity. A perhaps even greater challenge was to make the user respond to the new device in expected ways. Meters were broken into, coins were inserted on a string, and foreign coins would get stuck in the meter, leading one engineering report to soberly note that the meters could only be deployed in "localities where people were exceptionally honest".[14] In the initial stages of the invention, enormous trust had to be placed in the penny collector, who was also prone to help himself to coins from the piles he collected daily. But, in the end, for many engineers it was again technology that would solve these "human" problems. As one report put it, "both the penny collector and the householder may be dishonest; but, unlike the human heart, the metallic mechanism of these automata cannot be demoralised".[15]

In the early years, then, the problem of innovation in prepayment technology lay in gauging *how to distribute agencies.* Could measurement be left to the meter, or would the penny collector have to double-check? Could the collector be trusted not to serve himself? Would consumers be trustworthy or, as Clark might put it, "polite" enough not to tamper with the meter? The meter and the user thus needed to be co-constructed. Importantly, however, these co-constructions were inspired by and became allied with a liberal project. Indeed, it could be argued that the engineers' worries spoke to a larger liberal conundrum of distributing agencies and responsibilities: how to avoid governing too much, which population groups to include as reasonable members of the social contract and which to designate as in need of tutelage.

This continuous re-distribution of agencies, or what one might call *agencement* (Callon 2007), took the shape of an ongoing techno-political battle in which the technology itself became the terrain on which such ethical and political questions were negotiated. The autonomous self-governing citizen, so central to the liberal imagination, here emerges as a precarious socio-technical achievement, a figure unthinkable without the work of tools, devices, and infrastructures.

What Clark identified as British politeness, then, was not a foregone conclusion, but had to be painstakingly manufactured and was always prone to failure. And yet, Clark was correct in assuming that the meters would not simply travel smoothly to South Africa. The transfer of expertise, as Richard Rottenburg (2009) has shown, is predicated on a labor of translation. How a technology moves from context to context and what travels with it and what stays behind is thus an open question – once they leave their makers, technologies are "unstable objects" (Larkin 2008). And yet, as technologies move to new contexts, they also become re-stabilised: harnessed to new projects and anchored to new ethical regimes. It is this semiotic-material *work* that goes into making devices functional within specific ethico-political assemblages that I explore below.

Devices of counter-insurgency: "budget energy controllers" in late apartheid South Africa (Johannesburg, 1988)

As Clark himself was well aware, the context in which the prepaid meter came to be deployed in South Africa was quite different from late 19th century Britain. South Africa in the 1980s, he suggested, was "in a mess" because of "that political thing". For him, as for many other engineers I spoke to in the course of my fieldwork, talking about "politics" was not part of their job description and usually accompanied by frowns or pained expressions. Many had been in the business since the apartheid period and were at best uncertain about how to incorporate the anti-apartheid struggle into the narration of their professional biographies.

While the anti-apartheid struggle became known to the world outside South Africa through its campaigns for political rights, which for the most part were articulated at a national scale, it often took the shape of localised struggles that involved the more tangible, if less visible administrative connections to the apartheid state. During the "rent boycotts" in the 1980s, township residents all over South Africa withheld payment for rents and service charges as part of the effort to make the townships "ungovernable". Such acts of "fiscal disobedience" became both symbolic and material tools of insurgency with dramatic effects, disabling the running of township administrations and turning disconnections from services and evictions into sites of political struggle.[16] It was in this increasingly militant and militarised context of the boycotts that engineers began the search for technical solutions to the problem of non-payment.

At the same time, there were increasing moves towards the electrification of the townships by South Africa's powerful electricity parastatal Eskom. Over the course of the 1970s and 80s, it had become clear that "urban Africans" would not "return" to the Bantustans as grand apartheid ideologues had envisioned. This realisation prompted an increasing interest in the black urban population as an untapped market and as potential consumers of electricity and electrical appliances. It was in this profoundly paradoxical context of planned large-scale electrification and simultaneous politicised non-payment – of "reform" and counter-insurgency – that prepaid meters emerged.

As Peter Clark told me, in 1986 he developed the first South African prepaid meter. Much like in Britain, one of his primary tasks was to construct a functional assemblage of device, consumer and utility. Clark transformed the original technology primarily in two ways. First, in order to "protect" it from South African users, he replaced the coins with a magnetic card and non-transferable "tokens". Thus credit would be linked to a specific household and cash transaction would be limited to the local pay point. Secondly, the protective box around the meter needed to be made of cheaper, yet sturdy material to discourage residents from selling it as scrap metal. Apart from making changes in the technology, however, and similar to utilities in 19th century England, Eskom needed to configure its users.[17] One crucial issue, for example, was the creation of demand for electricity, a problem the utility sought to solve by providing residents with free basic appliances, such as lamps or hotplates. And yet, as several engineers told me, when visiting individual households at the time, they would often find the prepaid meter box blackened from the smoke emitted by a fire burning from a wood stove beneath the meter box, the electric hotplate not having been used at all.

By the late 1980s, prepaid meters commonly known at the time as "Budget Energy Controllers" were celebrated as "the next major technological breakthrough following the invention of the computer".[18] In many accounts, the political implications were stated bluntly. As one manufacturer suggested, prepayment "will help to depoliticise the supply of electricity such that energy does not become a pawn in the ideological struggles which the country is bound to face in the years ahead".[19]

The political target and effects of the invention became clearest in Clark's depiction of the early trials, during which engineers began simultaneous negotiations with the white municipalities and the rent boycotters in the adjacent black townships. In the late 1980s, Clark, by then employed by a meter manufacturer, started marketing the meters to individual municipalities *specifically* as devices that would enable the breaking of the boycotts and eliminate the need to access the townships directly to institute disconnections. Simultaneously, he marketed the meters to the boycott leaders, who desired electricity for the townships, but were unwilling to negotiate with the white municipalities. As Clark put it:

> We realised that we needed to go straight to the community. We … approached [the civic organisations] and said: "Look, we understand that you have a war, but we know you want electricity. With [prepayment], you will have both – electricity and no contract with the government. Nobody will ever come in again to disconnect you". There was a lot of psychology involved.

While Clark was eager to tell me about the successes, often such negotiations with the civics failed. Indeed, as one observer from the civics-aligned NGO Planact noted in 1992, prepaid meters were "fast becoming dangerously discredited" and had become "a source of real controversy in many townships, while some civics refuse to even discuss them" (Cobbett, 1992: 5f).

During the 1980s, prepaid meters were thus invested with the capacity to delink questions of payment and infrastructure from larger claims to citizenship and to re-establish and materially enforce the boundary between the administrative and the political. Thus, here the meters were harnessed to a late apartheid techno-politics that combined piecemeal "reforms" with a fierce defence of minority rule.[20] However, partly as a result of such contradictions and failures, many projects stayed at the trial phase and only relatively few prepaid meters were in fact deployed in the 1980s.

While the meters began their life as tools of moral improvement in the era of Victorian liberalism, in their move to late apartheid South Africa they were re-assembled as devices of counter-insurgency. In both moments, specific ethical and political projects were delegated to technology, and technology itself became a terrain on which such questions were expressed and negotiated. What joined the two moments, and what "traveled" with the meter, was the ability to delegate protracted ethico-political questions – of belonging, civic virtue, and indeed the limits of citizenship – to a technical terrain. The meters, as technical *forms* of "political society", thus produce what might be termed a graduated social contract by which citizenship is de facto mediated on an administrative terrain. And yet, even in a context of seemingly radical depoliticisation, such technical devices are constitutive of a material politics in a variety of ways and open to a diverse set of ethical claims and affective investments.

In contemporary South Africa, this material politics has taken several seemingly counter-intuitive forms. In a context of continued widespread non-payment in townships like Soweto and a neoliberal imperative for "cost recovery", prepayment technology has become the default mode of connecting poorer township residents to services, in the process often cutting previously unmetered access to services. Here, this graduated social contract maps onto the racial legacies of apartheid and is often experienced as punitive. Thus, as I elaborate in the next section, prepayment metering becomes the terrain for an ongoing techno-political struggle over the limits of citizenship. At the same time, the meters are at times deployed to *extend* services to residents of informal settlements. Here, the meters enable the connection of residents to the grid for the first time, whilst simultaneously rendering this connection precarious. Finally, there has also been an increasing *demand* for prepaid meters, often from wealthier residents who mistrust municipal accounting practices and see the meters as devices to wrest control from unaccountable city officials. Indeed, implicit in such demands for prepaid meters is often a desire to withdraw from the contractual relationship with a local state perceived as biased and untrustworthy. In an ironic reversal, this too produces a graduated social contract, if one initiated and demanded by residents themselves. Thus even in this context, the meter becomes object and terrain for ethical and political questions of trust, belonging and civic obligation. As prepaid meters increasingly become the default way to connect all residents to infrastructure,[21] yet other ways of assembling ethics, politics and technics may emerge.

Bad payers, smart meters and the market in anti-programs: the politics of "technical" innovation

After about an hour, my conversation with Peter Clark was interrupted by the Annual Prepayment Award ceremony that had been scheduled for that night. Sponsored by South Africa's largest cell phone company, the award is given each year to the company with the most innovative prepayment concept. The invention that won the award this year was a device called the Information Link at Point of Delivery (InfoPOD) developed by Peter Clark's employer, the multi-national company Actaris Metering Systems. The Minister of Minerals and Energy, Phumzile Mblambo-Ngcuka, had been invited to the ceremony to present the award. In her speech, she thanked Actaris for their tireless efforts for development and infrastructure in South Africa.

If at the award ceremony prepayment technology was presented as the transparent result of technological progress and South African ingenuity, throughout the conference this linear narrative had steadily unraveled. Earlier in the day, a group of protesters had gathered outside the convention centre armed with placards reading "Down with Prepaids!" This was not particularly surprising to the delegates who, after numerous demonstrations and a looming legal challenge, were well aware of the multiple objections to prepayment, especially for water. Meanwhile, inside the conference rooms, a less visible kind of politics emerged which involved the design of the technology itself. In the course of a host of PowerPoint presentations, what emerged was an industry in constant struggle with "non-technical" problems – government intrusions, legal hurdles, problems of standardisation and financing, and centrally, a variety of non-compliant consumers, who emerged as the protagonists in a seemingly perennial conflict over payment for services.

For despite the end of apartheid in 1994, many township residents had never resumed payment for services. Campaigns urging residents to pay for services in the name of national reconstruction had failed spectacularly, and in places like Soweto non-payment had in fact increased. Manual cut-offs from services and illegal reconnections were widespread, often aided by residents' organisations such as the Soweto Electricity Crisis Committee. At the same time, many municipalities began instituting neoliberal reforms and thus the pressure to institute "cost recovery" had grown dramatically. It was against this backdrop that prepayment technology came into wide use; this time, a technical solution to the failure of nationalist interpellation. In the ten years following the end of apartheid, five million households had thus been fitted with prepaid meters. Within a few years, however, many of these meters had been bypassed by residents, rendering them useless and giving residents illicit access to electricity. This in turn had left Eskom with the impossible task of manually checking and disconnecting each account. And it was *this* problem of the bypassed meters that preoccupied many presentations at the conference. I will focus here on two of the proposed solutions in order to illustrate this techno-politics of innovation.

As an arena of technology development, municipal engineering is in many respects unique in that it is addressed primarily at populations and not merely at individual users. Infrastructure developers are often centrally concerned with constructing and managing particular relations to utilities and the state, albeit relations usually conceived of as administrative rather than political. Indeed, in many of the presentations, what emerged most saliently were the relationships that could be established between the user and the technology and how these relationships would in turn mediate the population's relation to the utility or municipality. The *value* of the meters could only be established after an indication of how well it would fare in the establishment and durability of such relationships. This in turn required engineers to demonstrate not just knowledge of basic demographics, but also knowledge of a sociological kind, including a certain interpretive skill. For example, engineers needed to be aware of the history of payment practices in a particular area, which in turn was often bound up with the political histories of the townships. They needed to demonstrate how they had gone about testing the meters, which potential obstacles they had considered or what kinds of "social interventions" would be needed to ensure "user acceptability". Presentations thus often included the results of field trials or pilot projects that could demonstrate a certain *local* knowledge. The importance of this mobilisation of local knowledge became most obvious in the disjuncture between international and local presenters.

The day started with a presentation by a French representative of a large global infrastructure technology company who introduced the concept of "Automatic Meter Management" (AMM), a combination of prepayment and "smart" metering technology which would enable remote communication between the meter and the utility. The novelty of this technology was its claim to universal usability, that is, as the engineer put it, its ability to flexibly respond to "*global* challenges". In outlining each of these challenges, he created a peculiar map of the world from the perspective of the global trade in infrastructure technology that divided the world according to distinct "customer bases" with a diversity of needs, various levels of trustworthiness and sophistication, and a range of more or less predictable behaviour patterns. In the "first world", for example, the meters needed to be compatible with internet services to enable online billing or to offer customers flexible tariffs. In "third world" contexts and in places like South Africa, he argued, AMM would enable the "constant monitoring" of consumption and thus help eradicate bad payment, illegal connections and theft.

Standard metering, the engineer argued, was insecure and thus required constant auditing by utility employees. In a PowerPoint slide he summarised the resulting problems in the following way:

> "1000 Inspectors with 1000 dogs" can visit any account at least once a year…
> -If they were to know where to go…without political biases
> -If they are let in…
> -If this does not create political repulsion against the Utility…
> -Frauds can be reinstalled a day after the visit…
> -Evidence about frauds may be insufficient to recover past losses
> -…and why could not the readers execute this fraud eradication program before?
> -…and would anyone replace a dumb old electromechanical meter with a new, still dumb, electromechanical meter bound to be tampered with soon?

Older, "dumb" meters require a municipal official to audit the meter. This in turn opens the door for all sorts of problems. As the engineer suggested, residents may decide not to let him in, he may become the target for "political repulsion", or he may be unreliable himself. AMM would *delegate* the functions of auditing the meter from humans to technology.[22] Measuring, meter reading, surveillance and disconnections could now be performed by one technical device. Thus, the meter itself would "provide ways to bring [such] behaviours back to law and help dealing with the political issues of these customers". The technology would not only eliminate the unreliability of the municipal official and the possibility of bypassing the meter, but also the space for negotiations or protests left open by the presence of an official. Thus, the technology would, the engineer argued, perform "political" operations. AMM then was *one way* in which disputes over payment could be delegated to technology.

The presentation seemed to garner limited interest from South African delegates. It was quite clearly a costly solution that didn't speak to the financial constraints faced by South African utilities. More importantly, in the absence of evidence from a field trial, the engineer was unable to generate the *kind* of authority that would convince the other delegates. As one South African delegate argued, "Consumers will for some reason accept a technology type in one country, but in the next they will not accept it all. This behaviour is unpredictable and will ruin your venture and your investors' trust in your solution". What was required, in other words, were mechanisms that would limit the unpredictability of consumers' responses. And this, in turn, required a more sociological kind of knowledge and, indeed, a *thicker* description of the relations between communities and the local state.

It is here that the award-winning InfoPOD raised the interest of the other participants and quickly cast aside "global" solutions. In his presentation, the InfoPOD's developer, a local representative from Eskom, laid out the problem. Eskom could ill afford to replace its entire existing "dumb" prepaid meter base. On the other hand, many of the installed meters had been bypassed and had thus become useless. The InfoPOD, which had been piloted in the past year, would provide the solution. The InfoPOD – described by its developer as "a walk-by, non-intrusive, non-contact system to collect information from prepayment meters" – is a small radio device that can be attached to existing prepaid meters to enable officials to track tampering by accessing the meter remotely without needing physical access to residents' premises. Retro-fitted with the InfoPOD, the old meters would, as it were, become "smarter". Rather than having to enter the premises, utility employees could now drive or walk through the township with a radio receiver and automatically collect information from the meters at a distance.

Importantly, his presentation also included a discussion of the potential problems that could be expected during the introduction of the new devices. Asked how it would be possible to make people agree to such an installation, the engineer suggested: "When you come there, you don't only tell them that [the InfoPOD] is there to detect tampering. You tell them about the nice benefits. You have to be strategic, if I can put it that way". The success of such "technical" innovation thus relied centrally on engineers' local knowledge and on their capacity for thick description and strategic intervention.

Viewed from this perspective, the InfoPOD was only the latest instalment in a series of strategic measures to enforce payment – from appeals to civic virtue to the compulsions of a technical device – and thus to fashion a new relation between the local state and its citizens. It was an "anti-program", to use Latour's term.[23] What made this device valuable, then, was its capacity to strategically intervene in an ongoing conflict over payment, albeit one that was now carried out in the form of technology. It was just *one move* within an ongoing techno-political battle.

At the end of the conference, a delegate from Johannesburg told me epic stories about the various obstacles that had to be overcome by engineers throughout the history of prepayment technology in South Africa. Finally, he suggested, "you know the providers thought that people are uneducated, but in fact, a lot of innovation happens through them. If it wasn't for people regularly subverting the meter, we all wouldn't be here". And indeed, while much of this essay has been concerned with the experts at the conference, the absent protagonists were the "users", the "cunning water thieves", the "economic saboteurs", the "bad payers", "tamperers", "electricity poachers", and the residents with "political problems". Prepayment metering in South Africa is an industry that *requires* constant innovation, because the latest technologies are quickly out of date, having been pulled out, bypassed, broken, or re-wired. New anti-programs are thus required on a regular basis: "smarter meters", "security seals", tools that "audit", "track", "monitor", and "enable remote disconnections". Here, then, technical expertise is produced in constant conflict with what one might, with Michel Callon (Callon et al. 2009), term "expertise in the wild".[24]

Importantly, my point here is not to tell a story of small acts of resistance to rationalisation, neoliberalism or modernity; rather, my goal has been to map a techno-political terrain that often remains invisible in the common analytic focus on the public and on immaterial speech. Indeed, invested in the "technical" transformations I traced above are strategic scripts and counter-scripts – interventions in an ongoing series of low-intensity conflicts that have become materialised within the technology itself. Such a micro techno-politics concerns central moral-political questions about civic virtue and post-apartheid citizenship, albeit in a dramatically different *form*. In this context, infrastructure is not merely a tool or a symbol *for* the political, nor *merely* a conduit of power; infrastructure itself has become a political terrain. Indeed, this is a politics of non-publics, inscribed within pipes, wires, and technical devices as much as in the more visible protests that continue to make headline news on a regular basis in South Africa.

Conclusions

The South African story of the prepaid meter is of course in many ways specific. As the meter travels on, many different stories can surely be told. And yet, we might discover a similar *kind* of techno-politics in a diversity of locations and forms. There is today an unprecedented investment in developing sometimes highly sophisticated technologies for the poor and their presumed condition.[25] In much of the world, such technical devices increasingly mediate relationships between populations and the state or NGOs. Such technologies are often alternatively invested with a magical power to radically improve the lives of the poor or decried as tools of domination or surveillance. What I have suggested here is that seemingly neutral technical mediators come to do work within a diversity of ethico-political projects beyond their apparent pragmatism. Tracking the travels of such technical devices and ethnographically following the work of their inscription may thus enable us to "de-scribe" a politics in unfamiliar places and in unexpected forms. In turn, it might expand the conceptual and imaginative horizons of how we study and conceive of the political.

In a context in which the formal political sphere appears increasingly inaccessible, such material links often become the location at which political and ethical questions are negotiated and contested. Questions concerning citizenship, belonging, or civic virtue may here be expressed by flicking a switch, cutting off a wire or by installing a "smarter" meter. This is a politics far removed from the modern political imaginaries of a transparent, unencumbered sphere of public deliberation. Of course, at certain times, as during the protests in Chiawelo with which I began, this material politics does become public, transforming technics into a "matter of concern" (Latour 2004). But even here, looking more closely, "public" protest is often intimately tied to the more invisible forms of techno-politics I have outlined in this article.

Returning to the protest in Chiawelo, it turned out later that the protest was spawned, at least in part, by a new technical counter-script. Indeed, residents of Chiawelo had already had prepaid meters for a long time, but, as in other areas of Soweto, many of the meters had been bypassed. The protests, it became apparent, were prompted by a pilot project begun a few years earlier to install a new type of "split meter", an innovation which divides the meter in two parts: a touchpad to enter in the credit code located in the house and the actual meter now located outside on the pavements in unbreakable, "tamper-proof" green steel boxes – a new anti-program, one even more secure than the InfoPOD. And yet, three months after the protests, in October 2011, an audit by Eskom found that residents had opened the "unbreakable" green boxes with the help of grinders. Thus, the meters had been bypassed yet again, setting the stage for the development of new anti-programs in a seemingly endless cycle of innovation and subversion.

Notes

1 This chapter was previously published as: von Schnitzler, Antina (2013) Traveling Technologies: Infrastructure, Ethical Regimes, and the Materiality of Politics in South Africa. *Cultural Anthropology* 28(4): 670–693.

2 In 2007, out of ten million one-way prepaid meters deployed world-wide, eight million were located in South Africa. See Jones (2007).

3 See Stefan Helmreich's discussion of the concept of transduction (Helmreich 2007).

4 I am here inspired by Collier and Ong's (2005) account of the constitution of "global assemblages".

5 On the "user heuristic", see e.g. Fischer (1994), Bijker, Hughes and Pinch (1987). For compelling work on the technical and the political within scientific studies, see e.g. Akrich (1992), De Laet and Mol (2000), Hecht (2009; 2012) and Winner (1980). More recently, Michel Callon's account of "technical democracy" convincingly makes the case for a symmetry of traditional research and lay people's research (Callon et al. 2009). Yet, the *form* of political engagement he describes returns to more conventional notions of the political as located in a transparent, immaterial space of communication in which the rational, unencumbered individual human actor appears again as the protagonist.

6 This definition of the political as based on communication and the public is shared by a diverse array of theorists, from Arendt to Habermas and even more radical critics, such as Rancière.

7 See Madeleine Akrich's and Bruno Latour's work on this semiotic-material conception of inscription (Akrich 1992; Latour 1992).

8 For an analysis of the case, see Dugard (2010).

9 There is a large literature on new social movements in South Africa (see e.g. Naidoo 2007). Specifically on prepaid electricity see the edited volume *Electric Capitalism*, and especially the chapters by Ahmed Veriava and Prishani Naidoo, and by Peter van Heusden (MacDonald 2009). On the politics of the Treatment Action Campaign, see Robins (2008) and Comaroff (2007).

10 Despite the nationally mandated free provision of 50kWh per month, in 2004, poor Soweto households often spent 37% of their monthly income on basic services (Nefale 2004).

11 Following anthropological convention, I use pseudonyms for all individuals named in this article, apart from public figures.
12 Association of Gas Engineers and Managers, 1895, *Report of Proceedings 1895*, p. 302.
13 On the moral economies of measurement, see Gooday (2001) and Wise (1995).
14 "Penny-in-the-slot gas meters", Western Mail (Cardiff, Wales), Tuesday, November 8, 1898; Issue 9191.
15 "Penny-in-the-slot meters", Liverpool Mercury (Liverpool, England), Monday, June 19, 1893; Issue 14183.
16 I take the term "fiscal disobedience" from Janet Roitman (2005).
17 See Woolgar (1991) on the "configuration of the user".
18 G. Malan, October 1989, "Budget energy controllers can solve non-payment of water accounts" In: Municipal Engineer.
19 M.A. Stevenson, "Development of Prepayment Electricity Metering Systems for Use in First and Third World Environments", in Seventh International Conference on Metering Apparatus and Tariffs for Electricity Supply (Glasgow, Power Division, Institution of Electrical Engineers, 1992).
20 Ironically, it is the legacy of apartheid techno-politics that in part accounts for South Africa's contemporary status as a "global" leader in prepayment technologies.
21 In April 2013, City Power – Johannesburg's electricity utility – announced plans to install prepaid meters for all of its users.
22 On the concept of "delegation" see Latour (1992).
23 In coining the terms "programs of action" and "anti-program", Latour (1992) suggests that actions may be inscribed or anticipated within technical artifacts. These may however in turn be obstructed by other programs of actions, or anti-programs. Thus, an artifact can be "de-scribed" according to its programs of action (or anti-programs).
24 Callon (2009) uses the term "research in the wild" to refer to research by lay people outside the laboratory setting that often comes to challenge "secluded research" by professionals.
25 One might think here, for example, of food stamp cards in the US, biometric ATMs for rural populations in South Africa, informal banking systems, or the biometric registration of refugees in UN camps (see e.g. Breckenridge 2005; Elyachar 2012; Fassin 2011). See also Peter Redfield's work on humanitarian design (Redfield 2012).

References

Akrich, M., 1992, The de-scription of technical objects. In: W. Bijker and J. Law (eds), *Shaping Technology/Building Society: Studies in Sociotechnical Change*. pp. 205–224. Cambridge, MA: MIT Press.

Anand, N., 2012, Pressure: The politechnics of water supply in Mumbai. *Cultural Anthropology* 26(4): 542–564.

Appadurai, A., ed., 1986, *The Social Life of Things: Commodities in Cultural Perspective*. Cambridge: Cambridge University Press.

Arendt, H., 1990, *On Revolution*. London: Penguin.

Barry, A., 2001, *Political Machines: Governing a Technological Society*. London: Athlone Press.

Bennett, J., 2009, *Vibrant Matter: A Political Ecology of Things*. Durham: Duke University Press.

Bijker, W., Hughes, T.P. and Pinch, T., eds., 1987, *The Social Construction of Technological Systems: New Directions in the Sociology and History of Technology*. Cambridge, MA: MIT Press.

Braun, B. and Whatmore, S., eds., 2010, *Political Matter: Technoscience, Democracy, and Public Life.* Minneapolis: University of Minnesota Press.
Breckenridge, K., 2005, The biometric state: The promise and peril of digital government in the New South Africa. *Journal of Southern African Studies* 31(2): 267–282.
Callon, M., 2007, What does it mean to say economics is performative? In: D. MacKenzie, F. Muniesa, and L. Siu (eds) *Do Economists Make Markets? On the Performativity of Economics.* pp. 311–357. Princeton: Princeton University Press.
Callon, M., Lascoumes, P. and Yannick, B., 2009, *Acting in an Uncertain World: An Essay on Technical Democracy.* Cambridge, MA: MIT Press.
Chalfin, B., 2010, *Neoliberal Frontiers: An Ethnography of Sovereignty in West Africa.* Chicago: University of Chicago Press.
Chatterjee, P., 2004, *The Politics of the Governed: Reflections on Popular Politics in Most of the World.* New York: Columbia University Press.
Chatterjee, P., 2011, *Lineages of Political Society: Studies in Postcolonial Democracy.* New York: Columbia University Press.
Cobbett, W., 1992, Turning the meters back to zero: Electricity distribution and local government in Urban Areas. In: African National Congress (eds.) *Proceedings of the African National Congress National Meeting on Electrification.* Cape Town: University of Cape Town Press.
Collier, S.J., 2011, *Post-Soviet Social: Neoliberalism, Social Modernity, Biopolitics.* Princeton: Princeton University Press.
Collier, S.J. and Ong, A., eds., 2005, *Global Assemblages: Technology, Politics, and Ethics as Anthropological Problems.* Malden, MA: Blackwell Publishing.
Comaroff, J., 2007, Beyond the politics of bare life: Aids and the global order. *Public Culture* 19(1): 197–219.
Connolly, W.E., 2002, *Neuropolitics: Thinking, Culture, Speed.* Minneapolis: University of Minnesota Press.
Daston, L., and Galison, P., 2007, *Objectivity.* Cambridge, MA: MIT Press.
Daunton, M., 1983, *House and Home in the Victorian City: Working Class Housing, 1850–1914.* London: E. Arnold.
De Laet, M., and Mol, A., 2000, The Zimbabwe bush pump: Mechanics of a fluid technology. *Social Studies of Science* 30(2): 225–263.
Dugard, J., 2010, Legal mobilization in the struggle for water in Phiri, Soweto. In: A. Nielsen and S. Motta (eds) *Social Movements and the Development of Resistance in the Global South.* London: Palgrave.
Elyachar, J., 2012, Next practices: Knowledge, infrastructure, and public goods at the bottom of the pyramid. *Public Culture* 24(1 66): 109–129.
Fassin, D., 2011, Policing borders, producing boundaries: The governmentality of immigration in dark times. *Annual Review of Anthropology* 40(1): 213–226.
Ferguson, J., 2007, Formalities of poverty: Thinking about social assistance in neoliberal South Africa. *African Studies Review* 50(2): 71–86.
Fischer, C.S., 1994, *America Calling: A Social History of the Telephone to 1940.* Berkeley: University of California Press.
Gaonkar, D., and Povinelli, E., 2003, Technologies of public forms: Circulation, transfiguration, recognition. *Public Culture* 15(3): 385–397.
Gooday, G., 2004, *The Morals of Measurement: Accuracy, Irony, and Trust in Late Victorian Electrical Practice.* Cambridge: Cambridge University Press.
Hecht, G., 2009, *The Radiance of France: Nuclear Power and National Identity after World War II.* Cambridge, MA: MIT Press.

Hecht, G., 2012, *Being Nuclear: Africans and the Global Uranium Trade*. Cambridge, MA: MIT Press.

Helmreich, S., 2007, An anthropologist underwater: Immersive soundscapes, submarine cyborgs, and transductive ethnography. *American Ethnologist* 34(4): 621–641.

Hirschkind, C., 2001, Civic virtue and religious reason: An Islamic counterpublic. *Cultural Anthropology* 16(1): 3–34.

Jones, J.S., 2007, Prepayment reaches 21. *Metering International* 4(1): 80–81.

Joyce, P., 2003, *The Rule of Freedom: Liberalism and the Modern City*. London: Verso.

Kockelman, P., 2010, Enemies, parasites, and noise: How to take up residence in a system without becoming a term in it. *Journal of Linguistic Anthropology* 20(2): 406–421.

Larkin, B., 2008, *Signal and Noise: Media, Infrastructure, and Urban Culture in Nigeria*. Durham, NC: Duke University Press.

Latour, B., 1992, Where are the missing masses? The sociology of a few mundane artifacts. In: W. Bijker and J. Law (eds), *Shaping Technology/Building Society: Studies in Sociotechnical Change*. pp. 225–258. Cambridge, MA: MIT Press.

Latour, B., 2004, Why has critique run out of steam? From matters of fact to matters of concern. *Critical Inquiry* 30(2): 225–248.

MacDonald, D., ed., 2009, *Electric Capitalism: Recolonising Africa on the Power Grid*. Cape Town: HSRC Press.

Mahmood, S., 2005, *Politics of Piety: The Islamic Revival and the Feminist Subject*. Princeton: Princeton University Press.

Mains, D., 2012, Blackouts and progress: Privatization, infrastructure, and a developmentalist state in Jimma, Ethiopia. *Cultural Anthropology* 27(1): 3–27.

Mitchell, T., 2002, *Rule of Experts: Egypt, Techno-Politics, Modernity*. Berkeley: University of California Press.

Mitchell, T., 2011, *Carbon Democracy: Political Power in the Age of Oil*. London: Verso.

Naidoo, P., 2007, Struggles around the commodification of daily life in South Africa. *Review of African Political Economy* 34(111): 57–66.

Nefale, M., 2004, *A Survey on Attitudes towards Prepaid Meters in Soweto*. Johannesburg: Centre for Applied Legal Studies, University of the Witwatersrand.

Redfield, P., 2012, Bioexpectations: Life technologies as humanitarian goods. *Public Culture* 24(66): 157–184.

Robins, S., 2008, From "rights" to "ritual": Aids activism in South Africa. *American Anthropologist* 108(2): 312–323.

Roitman, J., 2005, *Fiscal Disobedience: An Anthropology of Economic Regulation in Central Africa*. Princeton: Princeton University Press.

Rottenburg, R., 2009, *Far-Fetched Facts: A Parable of Development Aid*. Cambridge, MA: MIT Press.

Star, S. L., 1999, The ethnography of infrastructure. *American Behavioral Scientist* 43(3): 377–391.

von Schnitzler, A., 2008, Citizenship prepaid: Water, calculability and techno-politics in South Africa. *Journal of Southern African Studies* 34(4): 899–917.

Winner, L., 1980, Do artifacts have politics? In: L. Winner (ed.), *The Whale and the Reactor: A Search for Limits in an Age of High Technology*. pp. 19–39. Chicago: University of Chicago Press.

Wise, M.N., ed., 1995, *The Values of Precision*. Princeton: Princeton University Press.

Woolgar, S., 1991, Configuring the user: The case of usability trials. In: J. Law (ed.), *A Sociology of Monsters: Essays on Power, Technology and Domination*. pp. 58–99. London: Routledge.

6 Water, housing and (in)formality in Kitwe, Zambia

Infrastructure, citizenship and urban belonging

Iva Peša

> We the residents of Wusakile constituency should not be treated as third class citizens. We are not happy with the manner water is supplied to our area. (The Post 2014)

Wusakile is one of the oldest and most prominent mining neighbourhoods in Kitwe, on the Zambian Copperbelt. Since the 1930s, mining companies have provided water and housing to their employees in Wusakile. Throughout the twentieth century Wusakile's residents enjoyed reliable running water, flush toilets and high-quality brick and mortar housing, in stark contrast to the informal areas of the city (Mutale 2004). This situation had changed quite markedly by 2014, when Wusakile "residents complained that they had had no water for many days, and even when there is water running in taps, it is rationed without any reason" (Zambian Watchdog 2014). In October 2014 widespread – and occasionally violent – protests erupted as a result. Erratic water supply sparked riots and brought up issues of equity, citizenship rights and belonging to the city. This is one occasion in which the relationship between infrastructure and citizenship gave rise to activism. More broadly, this chapter shows how "citizens' everyday access to, and use of, infrastructure in the city affect, and are affected by, their citizenship identity and practice" (Lemanski 2018). By exploring the variegated dynamics between water, housing and citizenship in Kitwe, this chapter highlights the ways in which citizens' access to water infrastructure is embedded in historical processes.

The vexed relationship between citizenship and infrastructure in Kitwe today has its roots in the historical dynamics of mining and urban governance on the Copperbelt (Fraser and Larmer 2010). Historical legacies have created a contested yet persistent distinction between "formal" and "informal" urban neighbourhoods. In Kitwe citizenship status depends on spatial and legal delimitations, distinguishing formal from informal neighbourhoods. Yet rather than denoting fixed categories, such distinctions must be understood as heterogeneous and hybrid, invoking a variety of state responses and popular practices (Varley 2013). This chapter argues that citizenship is materialised through access to infrastructure and that this informs citizens' sense of belonging to the city (Kangwa 2004). Infrastructure, in this sense, is highly political (Roy 2005). At the same time, infrastructure is always "in the making" as users, service providers and government officials struggle to keep the infrastructure going (Baptista 2018b). Through the inter-connected cases of water and housing this chapter explores the links between infrastructure, citizenship and urban governance in Kitwe (Peša 2015). It shows that "the policing of the arbitrary and fickle boundary between the legal and the illegal, formal and informal, is not just the province of the state but also becomes the work of citizens (…) [who] recreate the margins of legality and formality, imposing new socio-spatial differentiations", for example through varying access to water pipes or brick housing (Roy 2009: 85–86). In particular the chapter questions how "rights to the city" have been mediated through infrastructure (Harvey 2008). These "rights to the city" might either be interpreted as part of a human rights agenda of universal access to water and housing, or in a more radical Lefebvrian sense as a "right to define urban lived space" through appropriation and participation (Purcell 2003: 578). As the protests in Wusakile show, "rights to the city" are not merely future oriented, but are instead rooted in everyday experiences of infrastructure access. This chapter thus asserts the "usefulness of a historical and spatial perspective" in elucidating "the connections between past, present and future" (Baptista 2018a: 30).

How have "spatial differentiation and uneven development (…) been produced over time" and how has this become "embedded in the materiality" of infrastructure (Baptista 2018a: 33)? In the colonial era a binary formal–informal distinction determined citizens' right to access services in Kitwe. Yet in the contemporary neoliberal era of commercialised service provision, the boundaries between formal and informal access have blurred. This chapter explores these transitions in terms of power relations, citizen rights and water access. It thereby contributes to ongoing debates about infrastructure, the role of the state or the market in service provision, and how infrastructure influences citizenship (Bakker 2007; Pilo 2017). This chapter is based on a reading of published historical sources, grey literature on the water sector in Kitwe and qualitative interviews with water utility employees, water users and ordinary urban residents, conducted in 2014 and 2015. It situates current contestations over water access in longer term historical dynamics. This chapter first explores scholarly debates about water infrastructure, the role of the state and the market in water provision and how this influences citizenship. Thereafter, it examines the provision of water and housing in Kitwe from a historical perspective (from the 1940s until the present), revealing the importance and malleability of the formal–informal divide. Subsequently, the case of water kiosks is analysed in the context of the changing binary relationships between formal housing and de jure citizenship, and informal housing and de facto citizenship. In other words, how does access to water and housing inform experiences of citizenship amongst urban dwellers in Kitwe?

Infrastructure, urban governance and citizenship

Water infrastructure, due to its social, economic, political and cultural significance, in addition to its importance to public health, plays a pivotal role in urban life (Bakker 2014: 471). Water has accordingly been conceived of as a tool for improving living conditions within the right to services approach to inclusive urbanisation (Pilo 2017: 399). Yet calls for a universal right to water do not necessarily uproot unequal forms of water governance (Sultana and Loftus 2015: 97). Case studies show that "water is a brutal delineator of social power which has at various times worked to either foster greater urban cohesion or generate new forms of political conflict" (Gandy 2004: 363). Attention thus has to be paid to how urban water infrastructures distribute not only water, but also difference and inequality (Anand 2016).

Infrastructures "have played a significant role in processes of nation-building, modernization and development, as well as in the making of modern citizens" (Baptista 2018a: 32). Liberal-democratic citizenship, in particular, envisages a social contract between individuals and the state in which they consent to be ruled in exchange for access to services, such as water (Purcell 2003: 565). That is why "narratives of state and society have always been deeply entwined with the provision of water" (Angel and Loftus 2018: 2). From the 1850s onwards, the state has figured prominently as a water provider to its citizens. Bakker (2014: 471–472) explains how state-managed water provision improved living standards and facilitated social inclusion. Yet Gandy (2004: 374) reminds us that this relationship "is a fragile and historically specific phenomenon". Since the 1970s, neoliberal trends of privatisation of public services, such as water, have arguably eroded the power of the state, contributing to "dramatic changes in people's expectations of their citizenship rights" (Miraftab 2012: 191). Nonetheless, private sector involvement has not displaced the importance of the state, as governments continue to own, manage or regulate various aspects of water infrastructure (Bakker 2014: 487).

It therefore begs attention how power relations are being reshaped by neoliberal processes, notably the privatisation of urban services (Pilo 2017: 400). Neoliberal reforms assume that goods such as water will be more efficiently allocated through market forces, if treated as economic goods (Bakker 2007: 432). Yet water commercialisation has questionable distributional effects, as those who can pay for water have it readily but many others are left without affordable water access (Sultana and Loftus 2015: 98). Some authors have argued that the state does not have much power over processes of urban polarisation nor does it have the means to address widening inequalities in the distribution and quality of water supply (Gandy 2004: 372). Neoliberalisation processes, however, are fundamentally variegated (Baptista 2013: 591). There is evidence that privatisation has resulted in considerable government involvement in the water sector, especially in terms of regulation of the water market (Swyngedouw et al. 2002: 130–131). Shifting configurations of public–private partnerships have always been part of urban water supply (Swyngedouw et al. 2002: 126). Water infrastructure is, for instance, closely related to urban governance and municipal authorities play a crucial role in the roll-out of water to urban residents (Parnell and Pieterse 2010). In connection to this, the effect of neoliberal service reforms on urban citizenship has been debated extensively. Bakker (2014: 481) argues that neoliberalism has framed "consumers as customers rather than citizens" and that "consumer access is legitimated not by a citizen's entitlement to water as a service but rather by a customer's purchase of water as a quasi commodity". Similarly, Miraftab (2012: 191) asserts that neoliberalism has decentred "the state in the citizens' expectations of rights and well-being" by eroding public sphere responsibilities and delinking "citizens' civil, political and social rights".

Rather than eroding, however, the relationship between state power, urban citizenship and water access has been reconfigured. One example is the increasingly important role of human rights and other notions of rights (Purcell 2003). The "right to the city" and efforts to achieve a universal right to water are part of this trend (Purcell 2014). The realisation of such rights depends on effective state capacity to roll out urban services (Parnell and Pieterse 2010: 150). Sultana and Loftus (2015: 99) endorse the centrality of the state in enabling citizen rights. "Recognizing the right to water signals that authorities can be held politically and legally accountable, enabling those who are denied water to have means to contest and struggle for water". Even in case of fully commercialised neoliberal service provision, the state has a fundamental role to play in assuring equitable access to water as a human right and as part of urban citizenship. Because infrastructure shapes the daily practices of urban life, it equally informs particular expressions of citizenship. Pilo (2017: 398) usefully explores this "relationship between the 'right to the city' and urban services". This chapter shares Pilo's (2017: 411) interest in examining "the reshaping of power relations as part of a struggle to produce the city" by looking at how water infrastructure plays a role in defining people's "way of living in the city and the way they are recognised by the state and participate in society". By examining infrastructure from a historical perspective, this chapter explores how citizens "lay a claim in the present to a more just and fair urban future, one that situates urban dwellers as central in engaging in necessary socio-spatial change" (Silver 2014: 801). Instead of accepting a fixed distinction between "formal" citizens who can lay claim to individual piped water access and "informal" subjects who have to rely on shallow wells or streams (Gandy 2004: 368), the case of Kitwe suggests that attention should be paid to the ways in which "urban dwellers seek to adjust resource flows, reshape materialities, and experiment with multiple urban futures", illustrating their power to address urban inequalities (Silver 2014: 792). This aspect of negotiation and struggle reveals that "human rights are not necessarily sufficient criteria for the fulfilment of universal, equitable access to water supply" (Bakker 2014: 484). A struggle for the right to water "might mean more than simply achieving access to sufficient volumes of safe water. Potentially, such a struggle would mean achieving the right to be able to participate more democratically in the making of (…) the 'hydrosocial cycle'" (Sultana and Loftus 2015: 100). Examining how daily water access influences urban citizenship, belonging and relationships to the state therefore has much to say about desired and actual "rights to the city".

Through a case study of water provision in Kitwe, Zambia, this chapter shows that colonial patterns of infrastructure provision continue to matter today, as historical legacies complicate expectations of universal and undifferentiated service provision (Anand 2016). Water provision in Kitwe highlights the enduring role of private sector interests (mining companies and water utilities) in infrastructural development throughout the 20th century (Kazimbaya-Senkwe and Guy 2007). Historical studies are necessary to situate debates on state vs. market water provision in connection to citizenship. The persistent formal–informal divide in Kitwe has shaped water provision (household taps vs. shallow wells) and urban belonging. By looking at the interface between water infrastructure, urban power relations and citizen rights from a historical perspective, the possibilities and impossibilities of future pathways – including Lefebvrian notions of a "right to the city" – can be better understood. Neoliberal reforms and water commercialisation over the past two decades have not had predictable outcomes, and were not introduced onto a blank political or infrastructural canvas, but co-exist with previous and ongoing political and infrastructural interventions, mediated through colonial legacies of infrastructure provision, power relations and citizenship. The "ceaseless reconfiguration of urban [infrastructure] networks" is "an important site from which to analyze the sociomaterial production of cities" (Silver 2014: 788). By laying their own water pipes, for example, residents of informal areas claim a form of urban belonging that is both seceding from the state (in terms of installation) whilst simultaneously engaging with the state (by accessing public services) – whereby both acts demonstrate the rights and responsibilities of citizenship. Only by paying attention to such "historically and geographically specific material conditions" can simplistic binaries such as rights/commodities, formal/informal, public/private or citizen/customer be transcended (Bakker 2007: 434). This chapter seeks to understand how "rights to the city" have been mediated through infrastructure.

A historical perspective on (in)formality in Kitwe

In order to understand housing construction, infrastructure and service provision, it is necessary to appreciate "how much of African urban place making is rooted in historical and geographical relationships and reliant upon variant social relations with state power" (Myers 2010: 582). Kitwe's history as a mining town in many ways informs its current shape, even though urban built environments are rarely the outcome of straightforward historical trends (Fraser and Larmer 2010; Mususa 2012).

Kitwe was officially established as a mining town in 1935. By 1957 the city had a population of 86,000, which rose to 338,207 in 1990 and was estimated to be 522,092 in the 2010 census (Mutale 2004; CSO 2012). From the outset Kitwe had a dual character, divided between formal parts of the city, consisting of mining and city council settlements, and informal urban and peri-urban settlements. Informal settlements resulted "from unauthorised occupation of land, usually with non-adherence to land use and building regulations" (Huchzermeyer 2009: 59). In terms of infrastructure and service provision, the discrepancy between formal and informal areas has been and remains large (Kangwa 2004; Macmillan 2012). The discrepancy originated in the colonial period, but equally dominates post-colonial patterns of infrastructure and service provision (Larmer 2010). Of course the distinction between formality and informality is never clear-cut – not only are there differentiations within categories of informality, but the formal–informal divide itself is blurred (Roy 2005; Varley 2013; Baptista 2018b). Nonetheless, in Kitwe the distinction between formal and informal neighbourhoods – even if ever-shifting, arbitrary and fickle (Roy 2009) – has become inscribed in patterns of infrastructure provision. Whereas mining companies provided reliable services at highly subsidised rates to their employees (spacious brick housing, running water and electricity), these services were largely absent in informal settlements, where residents were excluded from public service provision, and consequently shallow wells, pit latrines and impermanent housing structures dominated (Kangwa 2004; Mutale 2004). City council areas, where civil servants and other formally employed residents lived, usually had access to water and electricity but not to the same standard as mining settlements – in council areas a water standpipe might be shared between several houses, whereas in mining areas each household had a tap (Mutale 2004). Nonetheless, there was a lot of variety within these categories. A 1959 report congratulated the mining industry for their "well-built, well-served and well-managed housing areas for African workers" (Regional Survey of the Copperbelt 1960: 77). The most luxurious three-bedroomed houses at the time would have a kitchen "fitted with a sink", a shower, as well as provisions "for power points in each room and also for an electric cooker in the kitchen" (Brown Report 1966: 58). Yet accommodation for single miners was often described as "unsatisfactory" or even "miserable" and particularly overcrowded (Moore 1948). Outside of mining areas, informal "squatter" housing was portrayed even more negatively as "unhygienic and squalid in the extreme" – although some houses in informal areas were commended for being "well-built and commodious" (Eccles Report 1944: 1, 13). This formal–informal divide, which linked notions of citizenship (based on labour status) to the provision of particular kinds of infrastructure, was rooted in colonial attempts at mobility and labour control (Macmillan 2012). In principle urban dwellers only had a right to water, electricity and housing (urban residence) as long as they were formally employed. What is more, only a very limited range of occupations (such as mining, nursing or civil service) were considered as legitimate employment, whilst construction workers and woodcutters, for example, were informalised or even banned from residing in the formal parts of the city (Larmer 2010). Through restrictive policies colonial rule purposefully sought to limit permanent urbanisation, especially before 1945, by encouraging temporary labour migration. Undeterred by these measures of control, increasing numbers of traders, charcoal burners and people simply seeking to access the alleged benefits of urban life, continued to move to and settle in informal urban areas (Macmillan 1993). Yet because residents of informal settlements lacked urban rights under colonial jurisdiction, they were cut off from public service provision. This created a contested infrastructural divide between formal and informal areas, which has persisted (albeit in an ever-shifting form) in the contemporary era (Mutale 2004).

After independence in 1964 movement control loosened and legal restrictions to urban residence were lifted. Because urbanisation rates increased rapidly, the demand for services outstripped supply and competition for housing, water and electricity access heightened. A formal–informal divide remained in place, as formal housing continued to be the preserve of mine employees and civil servants (Macmillan 1993; Fraser and Larmer 2010). Consequently, while informal settlements experienced inflated population numbers, service provision capacities did not grow in tandem (Kasongo and Tipple 1990). Mining companies, on the other hand, were "willing to continue to provide housing for their workers and to spend large sums in building new houses of improved standard" (Brown Report 1966: 56). Although mine housing in general "set a very high standard", some houses were still "below minimum standards" (Brown Report 1966: 57, 60). Mining companies considered themselves responsible "for the provision of electricity, water, street lighting and other community services", which former mineworkers today fondly recall as a "cradle to grave" policy (Brown Report 1966: 61; Interview with Edward Phiri, 18 August 2015). Yet such high-quality mine housing and infrastructural services could only be enjoyed by a minority of Kitwe's population. Mine housing remained inaccessible to the majority of casual migrants and the growing council waiting list for formal housing was never resolved (Macmillan 2012). Some government attempts to provide services to informal settlements did occur in the late 1960s and 1970s. Site-and-service schemes, part of a global approach led by the World Bank (Huchzermeyer 2009), were intended as a solution to growing urban population numbers and poor living conditions in informal settlements (Mutale 2004). In such schemes the council provided basic services, such as roads, standpipes, sewerage and electricity masts, whilst plot owners constructed their own houses and secured access to other amenities. However, progress on these schemes was extremely slow and only catered for a very limited number of informal residents (Kasongo and Tipple 1990). Between 1973 and 1982, Kasongo and Tipple (1990: 161) describe, "the population in squatter [informal] settlements increased three-fold in nine years, while houses increased to 218 per cent of the 1973 stock. At the same time, formal housing stock has increased hardly at all". Due to the inability of mining companies and local authorities to provide adequate housing and services for all urban residents, informal settlements continued to grow (Kasongo and Tipple 1990: 150). By 2010 the number of informal settlements in Kitwe had reached 35 and the gap between formal and informal areas in terms of infrastructure and service provision had widened (Interview with Irene Kayuni, 6 November 2014).

After a period of protracted growth and expansion since the late 1940s, the full impact of the collapse of copper prices hit the mining industry in Kitwe from 1975 onwards, resulting in a dramatic deterioration of living standards for most urban residents (Fraser and Lungu 2007; Mususa 2012). As employment levels on the mines declined, unemployment soared and the economy rapidly informalised, evidenced by activities such as street vending and urban agriculture (Ferguson 1999; Fraser and Larmer 2010). A severe polarisation in living conditions resulted from the economic crisis. Whereas those mineworkers who retained their jobs barely felt the economic crunch, residents of informal settlements had to queue for basic goods such as soap and cooking oil, while their houses were regularly threatened with demolition in an attempt by government officials to further exclude the informally employed and unemployed from urban rights and limited formal services (Kangwa 2004; Interview with Kabunda Kunka, 31 October 2014). Despite the generally bad performance of the mines, mineworkers continued to be a relatively privileged group, as interviewees still acknowledge:

> The mines continued with their "cradle to grave" policy and provided us with all our requirements. New houses continued to be built and we never experienced electricity or water shortages, everything was just the same as before. The real break came after the privatisation of the mines in 1991. (Interview with Ian Banda, 14 August 2015)

Following stringent Structural Adjustment Programmes, the mines were privatised from the 1990s onwards. In Kitwe, Nkana mine was purchased by Mopani Copper Mines PLC (a joint venture between Glencore International AG, First Quantum Minerals Ltd and ZCCM Investment Holdings) in 2000. Private mining companies were neither willing nor able to continue to provide costly social and infrastructural services to their employees (Fraser and Lungu 2007). As Mususa (2010: 381) explains, "the privatisation of the mines and the ensuing major downsizing of the workforce marked the start of an unprecedented economic crisis, with drastic effects on the local population". After privatisation, former mine housing was sold off to sitting tenants at reduced rates, whilst commercial utilities took over water services and started charging user fees (Kazimbaya-Senkwe and Guy 2007; Mususa 2012). On the Copperbelt, influenced by pressure from international financial institutions such as the World Bank, the water sector was commercialised, which entailed "separating the water services from the rest of municipal services, strict financial performance targets, operating water services on commercial principles, and the introduction of market-based remuneration for managers" (Chitonge 2010: 602). Commercial water utilities remained owned by local authorities, but operated as separate commercial entities – largely dependent on donor funding. In Kitwe, Nkana Water and Sewerage Company (NWSC) was established in 1998 and is jointly owned by Kitwe City Council and Kalulushi Municipal Council (www.nwsc.com.zm). Kazimbaya-Senkwe and Guy (2007: 879) explain how:

> The formation of these commercial utilities is premised on a fundamental neo-liberal argument in which it is envisaged that, separation of water functions from the local authorities, will bring political independence, whilst commercialisation will promote cost recovery, and some kind of private sector efficiency, all of which will lead to better services and connectivity.

This situation still shapes water and housing provision in Kitwe, influencing citizens' (im)possibilities of accessing infrastructural services.

Processes of privatisation and subsequent erratic service delivery (Fraser and Lungu 2007) have undermined the high-quality water, electricity and housing services formerly provided by the mines. In mining areas population densities rose and the population diversified, as ownership/occupation became based on affordability rather than employment status. Furthermore, retrenched mineworkers who had bought houses using their terminal benefits sought to make money by building backyard "cabins" to rent out or for business purposes (Mususa 2012). The subsequent increase in population size caused a deterioration in infrastructural services in mining settlements. Concurrently, some informal settlements have witnessed a construction boom and secured access to reliable water services (Mutale 2004). Informal settlements attracted a wide range of people, some of whom managed to gain considerable wealth through trade or business (Kasongo and Tipple 1990). Blurring formal–informal distinctions, some former mineworkers moved to informal settlements to avoid building regulations and/or because life is considered to be slightly cheaper there. Mususa (2010: 381) describes how some "moved up and some moved down" as "wide variations in the physical infrastructure of houses became clearly visible in the varying degrees of dilapidation and renovation, indicating horizontal inequalities among residents". From 2004 onwards a period of economic boom occurred, but this did not significantly improve service levels in Kitwe's mining areas (Fraser and Larmer 2010). Kazimbaya-Senkwe and Guy (2007: 877) describe illegal connections and water wastage, as "many households did not keep their installations and fittings in good working order". In contrast to infrastructural deterioration in formal areas, there have been numerous attempts to "upgrade" informal settlements, for example by providing water kiosks and by constructing low-cost housing with the help of NGOs (Peša 2015). Numerous residents of informal areas are building large houses, laying pipes for water connections and gaining access to the electricity grid for the first time (Interview with Innocent Kosamu, 16 October 2014). In some cases, water provision or housing in informal areas is now more reliable and of a higher standard than in historically formal settlements (Fraser and Larmer 2010). Nonetheless, the remainder of this chapter shows that the formal–informal divide (notwithstanding mutations) has proven remarkably resilient and that access to infrastructure continues to underpin inequalities in urban citizenship (Kangwa 2004).

The relationship between citizenship and infrastructure in Kitwe today can only be understood by appreciating these historical dynamics (Larmer 2010; Waters 2015). The water protests in Wusakile erupted as a result of a long legacy of mine provisioning and its discontinuation after privatisation (Interview with Leonard Kamanga, 28 October 2014). Although the outline of mining booms and busts, service provision and (in)formality might be familiar (Fraser and Larmer 2010), it has less frequently been examined how and why a division between formal and informal areas, especially in terms of infrastructure, has perpetuated after mine privatisation. Until 2015, no informal settlements were formally recognised by the Kitwe City Council and consequently all informal residents lacked land rights and even basic services (Kangwa 2004). Yet due to recent settlement upgrading and initiatives to arrange individual water connections, residents in certain informal settlements are claiming urban belonging (Interview with Kabunda Kunka, 31 October 2014). At the same time, mining employees who had long enjoyed privileged citizenship and infrastructural services have lost these as a direct result of privatisation (Mususa 2012). By looking at the case of water it can be examined how infrastructure and service provision have influenced citizenship in Kitwe. What are the "everyday processes through which users produce, maintain, and contest 'formal' and 'informal' infrastructures" in uneven urban infrastructural landscapes (Baptista 2018b: 2)? Although infrastructure can strengthen claims to citizenship, the development of infrastructure has been less than equitable and infrastructure alone is not enough to ensure "rights to the city" (Anand 2016).

Water kiosks: access to water for all?

Nkana Water and Sewerage Company is a commercial utility which started operations in 2000. It claims to be "committed to enhancing service provision and continuously providing adequate Water and Sewerage Services to all customers in Kitwe" (NWASCO 2013; www.nwsc.com.zm). Prior to its establishment, water in Kitwe was either provided through the mining companies or through the city council, at highly subsidised rates (Kazimbaya-Senkwe and Guy 2007; Dagdeviren 2008; Chitonge 2010). Yet only formal settlements, comprising mine employees and civil servants, were supplied through these channels (Kasongo and Tipple 1990; Mutale 2004). In informal settlements public water services were not provided and residents were expected to supply themselves, by resorting to shallow wells, illegal connections or the Kafue River (Self 2010; UN-Habitat 2012). Although NWSC is a commercial utility, government is the main shareholder (NWASCO 2013). Social considerations ("water as a right"; Bakker 2007; Sultana and Loftus 2015), international donor lobbying and government pressure have been pushing water provision into previously unserved informal and peri-urban areas (Robinson 2002). Hence, over the last decade NWSC has engaged in service expansion, in an attempt to "improve institutional capacities and operational capacities to match customer expectations for sustainable service delivery" (www.nwsc.com.zm). One manner in which they have done this is by expanding water provision to heavily populated informal and peri-urban settlements by means of water kiosks (Self 2010).

In an attempt to reverse the previous pattern of water provision, which had resulted in "exclusivist rather than inclusivist development", NWSC has engaged in establishing kiosks in informal and peri-urban areas since 2001 (Robinson 2002: 851). Kiosks aim to "deliver safe and reliable water at affordable prices to residents of low-income areas" (GTZ 2009: 2). Rhetorically at least, NWSC established kiosks not only to expand access to clean drinking water to previously unserved customers, but also to diminish the incidence of water-borne diseases, enhance school performance, worker productivity and even to stimulate economic growth (Robinson 2002). Providing water to low-income urban households through the low-cost technology of the water kiosk (simply described as "one-roomed buildings where piped water supply has been made available for the community to go and draw water from" (Interview with Mr Chenshe, 10 October 2014)) could, therefore, have pro-poor inclusive effects. The first kiosks in Kitwe were built in 2001 and since then their number has risen to 135, with each kiosk serving a prospective 1,000 households (NWASCO 2013). This increase in the number of kiosks, however, conceals some fundamental underlying problems, which threaten the inclusive potential of water provision to low-income urban areas (Robinson 2002; Self 2010).

Kiosks have "considerable potential" to include peri-urban and informal settlements in urban water provision networks (Robinson 2002: 856). Yet water kiosks faced problems from the outset and effective demand has remained low. It proved difficult to pick appropriate locations for the kiosks, vendors were not forthcoming and rumours plagued kiosk water, suggesting that drinking water from kiosks would cause barrenness (Self 2010; Interview with Mwangala Chiwala, 10 October 2014; Interview with Innocent Kosamu, 16 October 2014; Interview with Albert Malama, 21 October 2014). In some areas where alternative water sources are lacking, such as in Racecourse, customers might purchase water from the kiosks on a daily basis. Even so, customers whose houses are at a slight distance from a kiosk prefer alternative water sources such as shallow wells, illegal connections or nearby schools and clinics (Interview with Margaret, 22 October 2014). Consequently, some kiosks receive only a handful of customers a day, or might fall into disuse for weeks at a time. The cost of water does not seem to be a major deterring factor to kiosk use (Ashraf et al. 2007). Customers consider kiosk water to be affordable, although the cost of water per m^3; is slightly higher than that paid by households with a metered tap connection (Kazimbaya-Senkwe and Guy 2007; Self 2010; NWASCO 2013).[1] Customers are prepared to pay for water, as long as this enhances service delivery and reliability. This willingness to pay (Bakker 2014) is attested by the increasing number of individual tap connections, whereby individuals pay NWSC to lay pipes and connect their house to the main water supply network. Especially if a house is located at a distance from the existing water supply network, household tap connections require a high upfront investment. Yet even in peri-urban and informal areas where kiosks have recently been constructed, the number of individual tap connections is increasing and is challenging kiosk use (Interview with Innocent Kosamu, 16 October 2014). When assessing the use of kiosks, the NWSC community spokesperson concluded that: "The challenge we have is willingness to use, not affordability. It is about valuing the treated water, not really affordability" (Interview with Mwangala Chiwala, 10 October 2014). What has hampered the inclusive potential of water kiosks in Kitwe?

Rather than costs, the reasons for kiosk disuse seem to be related to patterns of urban governance, which create a division between formal and informal areas (Self 2010). Water kiosks have been erected in peri-urban and informal areas where previously no water services had been provided in any regulated manner (Kazimbaya-Senkwe and Guy 2007). Yet kiosks have not been distributed evenly across these areas, but have instead been placed only in those informal settlements which have been recognised for upgrading by the Kitwe City Council. Areas earmarked for upgrading have to meet a number of requirements set by the council, in terms of land tenure, planning and age of the settlement (Mutale 2004; Interview with Irene Kayuni, 6 November 2014). Such requirements favour those peri-urban and informal areas which already enjoy a certain degree of "formalisation", frequently hosting slightly better-off residents who have managed to build large permanent houses. Some of these households might already have invested in individual water connections (yard taps), by buying their own pipes and connecting to the main NWSC water network (Interview with Kabunda Kunka, 31 October 2014). Yard taps subsequently function as substitute kiosks, where neighbours can come to draw water at a fee to the tap proprietor. Whereas informal settlements recognised for upgrading might already have invested in alternative water infrastructure and therefore rarely use kiosks, Kitwe's formal–informal divide ironically excludes both unrecognised informal settlements and formal settlements (e.g. former mine areas) from accessing water kiosks (Kangwa 2004; Interview with Innocent Kosamu, 16 October 2014). The technical manager explained that NWSC is not willing to invest in areas which do not have security of tenure, because "you don't know when they [city council] will demolish it – and looking at the limited resources we have, I don't think it is wise to put them into an area that might be demolished" (Interview with Mr Chenshe, 10 October 2014).

Conceptualising water as a commodity rather than a right (Bakker 2014), this refusal of the commercial utility to serve households outside of its operational area means that settlements which have not been legalised are excluded from formal water services, such as kiosks. It is in these informal settlements, however, where the demand for water is greatest because residents lack alternative regulated water sources (Robinson 2002; Kazimbaya-Senkwe and Guy 2007). In the informal settlement of St. Anthony's, women and children can be found queuing at a metered tap minimally and reluctantly serviced by NWSC, where water is sold at prices well below kiosk fares because the water quality is sub-standard. The tap water is brown and contaminated, and consequently cholera is rife in the rainy season. Tap users in

St. Anthony's call for the establishment of water kiosks, where water supply would be regular and quality would be regulated, even if this would entail a rise in prices (Interview with Bernadette Mwaba, 24 October 2014). The commercial utility NWSC could considerably increase its coverage and water sales by expanding into informal settlements such as St. Anthony's. But it is unlikely that this will happen at present due to the formal–informal divide, which is underpinned by considerations of tenure security as well as the profitability of water sales (Robinson 2002; Kangwa 2004).

In principle, water kiosks hold the potential of including formerly excluded informal settlements into formal networks of urban water provision at relatively low costs (Robinson 2002). Yet in informal settlements recognised for upgrading the capacity of kiosks is not utilised, because residents prefer individual tap connections to the main line ("full access") or continue to use unregulated water sources such as shallow wells (Self 2010). On the other hand, residents of informal settlements which have not been provided with water kiosks might clamour for their establishment. In unrecognised informal settlements, such as St. Anthony's, the establishment of water kiosks would mark an improvement in formal and regulated water access, strengthening residents' claims to inclusion into the city (Interview with Bernadette Mwaba, 24 October 2014; Anand 2016). Water kiosks, in this instance, can become a first step to legitimate citizens' rights to the city and their inclusion into urban infrastructure networks (Kangwa 2004). However, where opportunities exist for individual tap connections households generally try to pursue these, even if it entails higher costs, as tap connections signal full service provision and are associated with stronger claims of citizenship (Kazimbaya-Senkwe and Guy 2007). Through their own initiative, by laying their own pipes, residents of informal settlements thus connect themselves to "formal" infrastructure networks and thereby seek recognition not only from the commercial service provider (NWSC) but also from the state (Kitwe City Council).

When considering water, citizenship and equity, the case of Wusakile provides a peculiar reversal of fortune. Due to the commercialisation of water services (from mining company to commercial utility) and population increase, water demand in Wusakile has come to outstrip supply potential. Old pipes and poorly maintained networks mean that water in some sections will only flow for one hour a day. Formerly well-served residents now have to walk long distances to access water or get water from unregulated sources. Although each household has an individual tap connection, no water is coming out of the pipes. Erratic water supply in Wusakile, making residents feel like "third class citizens", as the Member of Parliament phrased it, fuelled the riots in 2014. The establishment of kiosks is not viable in this case, as existing water connections are obstructing alternative water provision infrastructure (Interview with Leonard Kamanga, 28 October 2014). Only a complete (and very costly) rehabilitation of existing pipelines would restore reliable water access for Wusakile's residents, but despite intense lobbying by individuals, politicians and NGOs, plans for this have not advanced much by 2018 (Daily Nation 2018).

As these examples show, water provision is closely linked to ideas of citizenship, but different areas experience different dynamics (Swyngedouw 2009). Whereas informal settlements recognised for upgrading reject water kiosks and aspire to individual tap connections as a sign of full citizenship, unrecognised informal settlements regard water kiosks as a recognition of their belonging to the city. Wusakile provides a different case, as their previously reliable water connections have now fallen into disuse and users are resorting to unsafe water sources, which challenges their formerly privileged citizenship status (Larmer 2010). Water access in urban Kitwe is thus following skewed and unexpected courses, which at times confirm and at other times question notions of citizenship and belonging to the city by challenging and remaking the formal–informal divide (Waters 2015).

Negotiating (in)formality through infrastructure

Through the example of water kiosks, this chapter has sought to show how infrastructure can mediate notions of citizenship and urban belonging. By situating water kiosks in the historical context of water provisioning in Kitwe, the possibilities and limitations of more equal access to water infrastructure can be appraised. Debates on infrastructure have long discussed whether the state or the market would be more efficient and equitable in providing water services to citizens (Bakker 2007). It has been little highlighted, however, that "private sector participation is not new (…) but has been part and parcel of providing water services" throughout the twentieth century (Kazimbaya-Senkwe and Guy 2007: 870), especially on the Zambian Copperbelt where private mine involvement in water supply has played a role since the 1920s. Analysing the historical interdependencies between state and market provision of water obfuscates narratives which connect the commercialisation of infrastructural services to a weakening of citizen rights. Moreover, historical studies can confront the obduracy of infrastructure and the path-dependencies embedded in them (Baptista 2018a: 30). Kazimbaya-Senkwe and Guy (2007: 882) explain that: "The obduracy of water artefacts, the momentum of water technologies, and policy continuity means that, even today, some social groups in the Copperbelt are more predisposed to get access to water than others". On the Zambian Copperbelt, a formal–informal divide was historically rooted in and influenced by not only service provision but also claims to citizenship and urban belonging. It is important to bear in mind that categories such as "formal" and "informal" are always differentiated, embodying varying degrees of power and exclusion (Roy 2005). Although the formal–informal divide in Kitwe was indeed profound, it could nonetheless be reconfigured in distinct ways.

To illustrate such reconfiguration, Silver (2014) has introduced the concept of "incremental infrastructure". The laying of water pipes by informal residents in Kitwe would fit this concept, as "over time, previously informal dwellings develop the appearance and recognition of formal, robust structures. This solidifies not only the building but the household's status in the community and the perceived right to land tenure" (Silver 2014: 796). As Pilo (2017: 399) further explains:

> Often, public authorities deny people access to urban services by refusing to recognize them as city residents based on their illegal land tenure status. Thus, when inhabitants start to install and build infrastructure themselves – water pipes and electricity lines – they do so not only to improve their daily living conditions but also to demand recognition from the state through the construction of urban networks.

Infrastructure is thus intimately connected to state recognition and experiences of citizenship. Incremental acts such as the laying of water pipes or building a solid house in an informal area can be seen as "experiments in material configurations that seek to test and prefigure new forms of infrastructure and accompanying resource flows" (Silver 2014: 791). As Miraftab (2012: 195) argues, "for populations without access to legal or formal channels of citizen inclusion, informal practices offer an unnoticeable but effective means to assert their rights to a dignified livelihood". By looking at the case studies of water and housing in Kitwe, this chapter has examined the connections between infrastructure, urban governance and citizenship, the importance of historical legacies and the possibilities of future power reconfigurations.

Conclusion

Like most African cities, Kitwe has long been characterised using the binary analysis of formal–informal (Mutale 2004). As this chapter demonstrates, this formal–informal divide is not static, but has been continuously renegotiated and reconfigured throughout history, often mediated through access to infrastructure, such as housing and water (Larmer 2010; Mususa 2012). This chapter has traced the historical roots of Kitwe's formal–informal divide in terms of urban citizenship, as well as providing an example of a contemporary infrastructural project which has challenged yet equally remade the formal–informal divide. In Kitwe water kiosks have reconfirmed the formal–informal divide because upgraded settlements do not use them, whereas kiosks are not established in informal settlements (Self 2010). Claims of urban citizenship are thus played out through access (or lack of access) to water and housing. This chapter has questioned how infrastructure informs citizenship, rights to the city and feelings of urban belonging (Myers 2010). A study of infrastructure thus enhances understanding of "the relationship between development, materiality and social change in the city" (McFarlane 2008: 342). As it illustrates the possibilities and limitations of reconfiguring the formal–informal divide, a historical analysis of urban infrastructural power relations is imperative to imagining more equitable futures.

Although infrastructure is important for making claims of belonging to the city, infrastructure alone cannot remake the historical formal–informal divide. The history of infrastructural development in Kitwe until now has been largely unequal and exclusionary, rather than inclusive or equitable (Mutale 2004). Neither state nor market provisioning of infrastructural services has ensured water access for all (Bakker 2007). Whilst legacies of colonialism and mining enforced a formal–informal divide, the neoliberal blurring of access to services has not eliminated this divide or its inherent inequalities. Considering issues of urban governance is crucial in this respect, because "infrastructures are socio-technical in nature, a wide range of social, material, technical, political, legal and economic factors come to the fore" (McFarlane 2008: 348). In order to craft more equitable infrastructural policies that create a broader basis for inclusive forms of infrastructural citizenship, it is paramount to understand the history of infrastructure and (in)equality in urban areas. The situation in Wusakile and the apparent reversal of fortune in terms of water provision in the area can only be comprehended by considering the history of mine provisioning and the privileged status of mining settlements vis-à-vis informal settlements (Larmer 2010; Macmillan 2012). Water kiosks can challenge and reconfigure the formal–informal divide, but infrastructure alone is not sufficient to substantiate urban rights. To a certain extent, infrastructure can remake the city and create new models of urbanisation (Harvey 2008), but historical trends and the materiality of infrastructure itself have equally propelled the continuation of existing patterns of urban inequality, in the case of Kitwe expressed through a formal–informal divide.

Note

1 For current water tariffs in Kitwe see: www.nwasco.org.zm/files/WATER_AND_SEWERAGE_TARIFFS_FOR_2017_final.pdf (accessed 10 January 2018).

References

Anand, N., 2016. Hydraulic Publics. *Limn*, 7. https://limn.it/articles/hydraulic-publics/ (accessed 27 November 2018).

Angel, J. and Loftus, A., 2018. With-against-and-beyond the Human Right to Water. *Geoforum*. doi: doi:10.1016/j.geoforum.2017.05.002

Ashraf, N., Berry, J. and Shapiro, J.M., 2007. Can Higher Prices Stimulate Product Use? Evidence from a Field Experiment in Zambia. NBER Working Paper, 13247. www.nber.org/papers/w13247 (accessed 27 November 2018).

Bakker, K., 2007. The "Commons" Versus the "Commodity": Alter-globalization, Anti-privatization and the Human Right to Water in the Global South. *Antipode*, 39(3): 430–455.

Bakker, K., 2014. The Business of Water: Market Environmentalism in the Water Sector. *Annual Review of Environment and Resources*, 39: 469–494.

Baptista, I., 2013. The Travels of Critiques of Neoliberalism: Urban Experiences from the "Borderlands". *Urban Geography*, 34(5): 590–611.

Baptista, I., 2018a. Space and Energy Transitions in Sub-Saharan Africa: Understated Historical Connections. *Energy Research & Social Science*, 36: 30–35.

Baptista, I., 2018b. Electricity Services Always in the Making: Informality and the Work of Infrastructure Maintenance and Repair in an African City. *Urban Studies*, 1–16.

Chitonge, H., 2010. Who is Subsidising Whom? Water Supply Cross-subsidisation Policy, Practice and Lessons from Zambia. *The Journal of Modern African Studies*, 48(4): 599–625.

CSO, 2012. *Zambia 2010 Census of Population and Housing: Volume 11 National Descriptive Tables*. Central Statistical Office: Lusaka. http://journals.plos.org/plosone/article/file?type=supplementary&id=info:doi/10.1371/journal.pone.0141689.s006 (accessed 10 January 2018).

Dagdeviren, H., 2008. Waiting for Miracles: The Commercialization of Urban Water Services in Zambia. *Development and Change*, 39(1): 101–121.

Daily Nation, 2018. Wusakile MP Pledges k100,000 for School Project, *Daily Nation*, 9 April. www.dailynation.info/wusakile-mp-pledges-k100000-for-school-project/ (accessed 5 September 2018).

Eccles, L.W.G., 1944. *Report of the Commission Appointed to Inquire into the Administration and Finances of Native Locations in Urban Areas, Northern Rhodesia*. Government Printer: Lusaka.

Ferguson, J., 1999. *Expectations of Modernity: Myths and Meanings of Urban Life on the Zambian Copperbelt*. University of California Press: Berkeley.

Fraser, A. and Larmer, M. (Eds.), 2010. *Zambia, Mining, and Neoliberalism: Boom and Bust on the Globalized Copperbelt*. Palgrave Macmillan: New York.

Fraser, A. and Lungu, J., 2007. *For Whom the Windfalls? Winners & Losers in the Privatisation of Zambia's Copper Mines*. Civil Society Trade Network of Zambia: Lusaka.

Gandy, M., 2004. Rethinking Urban Metabolism: Water, Space and the Modern City. *City*, 8(3): 363–379.

Government of Zambia, 1960. *First Report on a Regional Survey of the Copperbelt 1959*. Government Printer: Lusaka.

Government of Zambia, 1966. *Report of the Commission of Inquiry into the Mining Industry, Brown Report*, Government Printer: Lusaka.

GTZ (Deutsche Gesselschaft für Technische Zussamenarbeit), 2009. *Case Study: Water Kiosks*. www.giz.de/expertise/downloads/gtz2009-0193en-water-kiosks.pdf (accessed 1 January 2018).

Harvey, D., 2008. The Right to the City. *New Left Review*, 53. https://newleftreview.org/II/53/david-harvey-the-right-to-the-city (accessed 27 November 2018).

Huchzermeyer, M., 2009. The Struggle for In Situ Upgrading of Informal Settlements: A Reflection on Cases in Gauteng. *Development Southern Africa*, 26(1): 59–73.

Kangwa, J., 2004. Zambia: "Having a Place of your Own" in Kitwe. In Home, R., and Lim, H. (Eds.), *Demystifying the Mystery of Capital: Land Tenure and Poverty in Africa and the Caribbean*. Cavendish: London, 121–143.

Kasongo, B.A. and Tipple, A.G., 1990. An Analysis of Policy towards Squatters in Kitwe, Zambia. *Third World Planning Review*, 12(2): 147–165.

Kazimbaya-Senkwe, B.M. and Guy, S.C., 2007. Back to the Future: Privatisation and the Domestication of Water in the Copperbelt Province of Zambia, 1900–2000. *Geoforum*, 38(5): 869–885.

Larmer, M., 2010. Historical Perspectives on Zambia's Mining Booms and Busts. In Fraser, A. and Larmer, M. (Eds.), *Zambia, Mining, and Neoliberalism: Boom and Bust on the Globalized Copperbelt*. Palgrave Macmillan: New York, 31–58.

Lemanski, C., 2018. Infrastructural Citizenship: Spaces of Living in Cape Town, South Africa. In Jones, A., Miller, B., Ward, K. and Wilson, D. (Eds.), *The Routledge Handbook on Spaces of Urban Politics*. Routledge: London, 350–360.

Macmillan, H., 1993. The Historiography of Transition on the Zambian Copperbelt – Another View. *Journal of Southern African Studies*, 19(4): 681–712.

Macmillan, H., 2012. Mining, Housing and Welfare in South Africa and Zambia: An Historical Perspective. *Journal of Contemporary African Studies*, 30(4): 539–550.

McFarlane, C., 2008. Urban Shadows: Materiality, the "Southern City" and Urban Theory. *Geography Compass*, 2(2): 340–358.

Miraftab, F., 2012. Right to the City and the Quiet Appropriations of Local Space in the Heartland. In Smith, M.P. and McQuarrie, M. (Eds.), *Remaking Urban Citizenship: Organizations, Institutions, and the Right to the City*. Transaction Publishers: New Brunswick, 191–203.

Moore, R.J.B., 1948. *These African Copper Miners: A Study of the Industrial Revolution in Northern Rhodesia, With Principal Reference to the Copper Mining Industry*. Livingstone Press: London.

Mususa, P., 2010. "Getting By": Life on the Copperbelt after the Privatisation of the Zambia Consolidated Copper Mines. *Social Dynamics*, 36(2): 380–394.

Mususa, P., 2012. Mining, Welfare and Urbanisation: The Wavering Urban Character of Zambia's Copperbelt. *Journal of Contemporary African Studies*, 30(4): 571–587.

Mutale, E., 2004. *The Management of Urban Development in Zambia*. Ashgate: Aldershot.

Myers, G.A., 2010. The Social Construction of Peri-urban Places and Alternative Planning in Zanzibar. *African Affairs*, 109(437): 575–595.

NWASCO (National Water Supply and Sanitation Council), 2013. *Urban and Peri-urban Water Supply and Sanitation Sector Report 2013*. Government Printer: Lusaka.

Parnell, S. and Pieterse, E., 2010. The "Right to the City": Institutional Imperatives of a Developmental State. *International Journal of Urban and Regional Research*, 34(1): 146–162.

Peša, I., 2015. Homegrown or Imported? Frugal Innovation and Local Economic Development in Zambia. *Southern African Journal of Policy and Development*, 1(2): 15–25.

Pilo, F., 2017. A Socio-technical Perspective to the Right to the City: Regularizing Electricity Access in Rio de Janeiro's Favelas. *International Journal of Urban and Regional Research*, 41(3): 396–413.

Purcell, M., 2003. Citizenship and the Right to the Global City: Reimagining the Capitalist World Order. *International Journal of Urban and Regional Research*, 27(3): 564–590.

Purcell, M., 2014. Possible Worlds: Henri Lefebvre and the Right to the City. *Journal of Urban Affairs*, 36(1): 141–154.

Robinson, P.B., 2002. "All for Some": Water Inequity in Zambia and Zimbabwe. *Physics and Chemistry of the Earth*, 27: 851–857.

Roy, A., 2005. Urban Informality: Toward an Epistemology of Planning. *Journal of the American Planning Association*, 71(2): 147–158.

Roy, A., 2009. Why India Cannot Plan its Cities: Informality, Insurgence and the Idiom of Urbanization. *Planning Theory*, 8(1): 76–87.

Self, J.A., 2010. *The World Bank's Framework for Service Provision: The Case of the Water Kiosks of the Zambian Commercialised Utilities* (MA Thesis). Saint Mary's University: Halifax, Nova Scotia, Canada.

Silver, J., 2014. Incremental Infrastructures: Material Improvisation and Social Collaboration across Post-colonial Accra. *Urban Geography*, 35(6): 788–804.
Sultana, F. and Loftus, A., 2015. The Human Right to Water: Critiques and Condition of Possibility. *WIREs Water*, 2(2): 97–105.
Swyngedouw, E., 2009. The Political Economy and Political Ecology of the Hydro-social Cycle. *Journal of Contemporary Water Research and Education*, 142: 56–60.
Swyngedouw, E., Kaïka, M. and Castro, E., 2002. Urban Water: A Political-ecology Perspective. *Built Environment*, 28(2): 124–137.
The Post, 2014, Wusakile MP Richard Musukwa. *The Post*, 27 October. http://postzambia.com/news.php?id=3351 (accessed 20 October 2015).
UN-Habitat, 2012. *Zambia Urban Housing Sector Profile*. UN-Habitat: Nairobi, Kenya.
Varley, A., 2013. Postcolonialising Informality? *Environment and Planning D: Society and Space*, 31: 4–22.
Waters, H., 2015. (Dis)connecting the Flow, Steering the Waters: Building Hegemonies and "Private Water" in Zambia, 1930s to the Present. In Harris, L.M., Goldin, J.A., and Sneddon, C. (Eds.), *Contemporary Water Governance in the Global South: Scarcity, Marketization and Participation*. Routledge: London, 133–148.
Zambian Watchdog, 2014. Wusakile Residents Fight with Police over Water Blues. *Zambian Watchdog*, 11 Octoberwww.zambianwatchdog.com/wusakile-residents-fight-with-police-over-water-blues/comment-page-1/ (accessed 10 January 2018).

List of interviews

All interviews have been conducted by the author. I would like to sincerely thank Lyness Mumba Lubemba, Maria Kankondo and David Phiri for their assistance in conducting these interviews and translating from Bemba to English.Interview with Mr Chenshe, NWSC Technical Manager, Kitwe, 10 October 2014.
Interview with Mwangala Chiwala, NWSC Community Relations Manager, Kitwe, 10 October 2014.
Interview with Innocent Kosamu, Kiosk Manager, Ipusukilo, Kitwe, 16 October 2014.
Interview with Albert Malama, Copperbelt University Lecturer School of Built Environment, Kitwe, 21 October 2014.
Interview with Margaret, Kiosk Manager, Racecourse Compound, Kitwe, 22 October 2014.
Interview with Bernadette Mwaba, St. Anthony's Compound, Kitwe, 24 October 2014.
Interview with Leonard Kamanga, Wusakile, Kitwe, 28 October 2014.
Interview with Kabunda Kunka, Settlement UpgradingUnit, Kitwe City Council, Kitwe, 31 October 2014.
Interview with Irene Kayuni, Head of Housing Department, Kitwe City Council, Kitwe, 6 November 2014.
Interview with Ian Banda, Wusakile, Kitwe, 14 August 2015.
Interview with Edward Phiri, Mindolo, Kitwe, 18 August 2015.

Conclusion

Infrastructures of citizenship

Charlotte Lemanski

Infrastructures of citizenship is a broad rather than prescriptive analytical framework, and can be defined in multiple ways. The aim of this book is not to re-define either infrastructure or citizenship, or even to create a narrow definition for infrastructures of citizenship, but rather to highlight the illogicality that while infrastructure and citizenship are arguably interlinked in terms of the functioning of the city, the two concepts are rarely considered collectively in theory.

As this book has demonstrated, urban infrastructure is a socio-technical process that reflects and perpetuates the inequalities of citizenship, while the relationship between citizens and the state (i.e. citizenship) is primarily mediated through infrastructure. Indeed, for most low-income urban dwellers (particularly in the global South), the primary way in which they "see" the state in their everyday lives is via access to (and the quality of) public infrastructure. While the most visible demonstrations of citizenship may emerge from collective radical protests (often over access to infrastructure), citizenship is also practiced in the everyday ways in which urban dwellers use, adapt, repair, and bypass forms of public infrastructure. Indeed, infrastructure is not exclusively provided by the state, but also by citizens themselves (as well as by private and NGO actors). In this sense, the ways in which citizens secure initial and long-term access to infrastructure is a form of citizenship *in practice*, that is regularly re-negotiated in dialogue with technical (e.g. taps, toilets, wires) and social (e.g. neighbours, engineers, and planning officials) forms of infrastructure. At the same time, infrastructures of citizenship also manifest in the civic *identities* of urban dwellers, whereby the quality of infrastructure that is accessible for an individual, household or community represents a reflection and perpetuation of unequal citizenship status in the city.

From the perspective of the state, the language of infrastructure has become a dominant theme of urban governance, through which (particularly low-income) urban dwellers are framed primarily as infrastructure consumers and/or complainers. Within this language of the state, urban dwellers' infrastructure demands and actions are rarely embraced as a reflection of active citizenship, but rather as representations of radical and/or illegal behaviour. And yet, despite these glaringly obvious connections between infrastructure and citizenship in terms of the ways in which the city is lived and governed in terms of people, places and processes, the two concepts and empirical practices are rarely explored in tandem. Consequently, this book comprises a short collection of scholars' reflections on the implications of their research on infrastructure for debates on citizenship (and vice versa), as an opportunity to highlight the ways in which infrastructure and citizenship are already intertwined in the city, and would benefit from greater connection within scholarly debates.

Each chapter interprets both infrastructure and citizenship in nuanced ways, situated according to the empirical location and sectoral focus of their study as well as the authors' epistemological framing. While Silver and McFarlane's joint chapter examines infrastructure in terms of the social connections that bind together (and at times, restrict) fellow residents of an informal settlement, in the other chapters infrastructure is primarily explored in terms of physical access to basic services such as housing (Lemanski, Wafer, Peša), water (Peša), electricity (von Schnitzler) and sanitation (McFarlane). These diverse interpretations and analyses of "urban infrastructure" are by no means in conflict, but provide complementary approaches that enrich the breadth of understanding, and reflect the situated nature of any analysis of urban life. The chapters each offer rich empirical findings based on case studies from a range of global South cities (from Kampala to Mumbai, Cape Town, Johannesburg and Kitwe), and while each narrative produces diverse interpretations of the connections between infrastructure and citizenship, there are two core themes.

The most common thread that emerges from the chapters is that of the impacts of *structural inequalities*: primarily, the ways in which infrastructure reflects and perpetuates existing urban fragmentation and division that is not merely technical-physical, but also inherently socio-political in mirroring and extending inequalities of citizenship status (in practice if not in law). The intertwined nature of these infrastructure-citizenship inequalities are particularly pertinent in Antina von Schnitzler's chapter on pre-paid electricity meters in South Africa. Her chapter demonstrates the physical infrastructural inequalities of the metering system whereby pre-paid electricity meters have been installed in low-income parts of Johannesburg, while wealthy suburban areas are billed monthly in arrears. This differentiated system of infrastructural monitoring and payment is not merely technical, but demonstrates a form of two-tier citizenship, where the state (in this example, via the public utility company Eskom) provides a differentiated type of service for different citizens despite legal equality for all. This is particularly problematic in South Africa, where the legacies of apartheid continue to dominate the urban landscape, and consequently the pre-paid meter is conceptualised as a tool of post-apartheid oppression. Unsurprisingly, low-income citizens seek to avoid payment and bypass the meter, and this infrastructural struggle is essentially a citizenship struggle, in which citizens are demanding their right to basic services in the context of state failure.

Examples of structural inequality in jointly determining experiences and practices of infrastructure and citizenship are equally dominant in other chapters: for example, Iva Peša's historical account of urban dwellers' access to housing and water in Zambia demonstrates a two-tier citizenship between mining employees and informal dwellers that was reliant on a simplistic formal/informal housing binary to justify differentiated citizenship rights and infrastructural access. Similarly, while Jonathan Silver and Colin McFarlane's chapter adopts a social perspective to infrastructure – demonstrating the ways in which infrastructure is not just about the materiality of pipes, bricks and concrete, but equally about the relationships between people that enable urban dwellers to function within the margins of the city – their analysis equally reveals the embedded nature of structural inequalities for both infrastructure and citizenship. Essentially, social infrastructure enables urban dwellers to extend beyond the structural inequalities of their "received" citizenship (as residents of technically "illegal" informal settlements), to forge a form of active citizenship that whilst relatively insular in outlook (focused on survival in the settlement) is communal in the forms of support offered between residents, creating a new form of citizenship that challenges structural inequalities.

The other core theme that emerged through a number of the chapters is the role played by *protest*, as a demonstration of citizenship that takes places in, and over access to, infrastructure. As discussed in the opening chapter by Charlotte Lemanski, radical protest is often the primary way in which infrastructure and citizenship literatures are connected, typically resulting in a failure to consider other less confrontational connections. However, what emerges from the book is that protest can function in both everyday and spectacular ways. For example, bypassing the electricity meter (von Schnitzler) is as much a form of everyday protest as violence (Wafer, Peša) and the threat of urinating in public spaces (McFarlance), which represent more spectacular forms of protest. Furthermore, the aim of infrastructure-based protest can be about demanding greater state involvement (McFarlane, Peša), campaigning for less state involvement (Wafer), or engaging in forms of protest that seek to undermine the state (von Schnitzler).

To conclude, infrastructures of citizenship represent an attempt to demonstrate the inter-connections in both theory and practice between the literatures and empirical trends of infrastructure and citizenship. In doing so, our understanding of both concepts and practices is expanded, as well as providing a richer analysis of city life. Inherent to both infrastructure and citizenship is the role of politics, both in terms of state action (official policies, unofficial practices) and citizen actions (everyday and spectacular) in determining where infrastructure exists and who can access it, as well as how it is used and maintained. The politics of resource allocation (where citizenship is as much a resource as infrastructure) emerges in all of the chapters, whereby the inequality of infrastructural provision reflects and perpetuates the inequalities of citizenship practices and identities. Essentially, this book argues that exploring citizenship through an infrastructural lens, and infrastructure through a citizenship lens, illuminates not only each concept/practice, but also provides the means for scholars, activities and policymakers to more effectively understand, plan and govern city life.

Index

Note: Entries in *italics* refer to *Figures*, and entries in **bold** refer to **Tables**.